Global Perspectives on Health Geography

Series editor

Valorie Crooks, Department of Geography, Simon Fraser University,
Burnaby, BC, Canada

Global Perspectives on Health Geography showcases cutting-edge health geography research that addresses pressing, contemporary aspects of the health-place interface. The bi-directional influence between health and place has been acknowledged for centuries, and understanding traditional and contemporary aspects of this connection is at the core of the discipline of health geography. Health geographers, for example, have: shown the complex ways in which places influence and directly impact our health; documented how and why we seek specific spaces to improve our wellbeing; and revealed how policies and practices across multiple scales affect health care delivery and receipt.

The series publishes a comprehensive portfolio of monographs and edited volumes that document the latest research in this important discipline. Proposals are accepted across a broad and ever-developing swath of topics as diverse as the discipline of health geography itself, including transnational health mobilities, experiential accounts of health and wellbeing, global-local health policies and practices, mHealth, environmental health (in)equity, theoretical approaches, and emerging spatial technologies as they relate to health and health services. Volumes in this series draw forth new methods, ways of thinking, and approaches to examining spatial and place-based aspects of health and health care across scales. They also weave together connections between health geography and other health and social science disciplines, and in doing so highlight the importance of spatial thinking.

Dr. Valorie Crooks (Simon Fraser University, crooks@sfu.ca) is the Series Editor of Global Perspectives on Health Geography. An author/editor questionnaire and book proposal form can be obtained from Publishing Editor Zachary Romano (zachary.romano@springer.com).

More information about this series at http://www.springer.com/series/15801

Prestige Tatenda Makanga
Editor

Practicing Health Geography

The African Context

Springer

Editor
Prestige Tatenda Makanga
Place Alert Labs
Department of Surveying and Geomatics
Midlands State University
Gweru, Zimbabwe

ISSN 2522-8005 ISSN 2522-8013 (electronic)
Global Perspectives on Health Geography
ISBN 978-3-030-63470-4 ISBN 978-3-030-63471-1 (eBook)
https://doi.org/10.1007/978-3-030-63471-1

This Springer imprint is published by the registered company Springer Nature Switzerland AG
The registered company address is: Gewerbestrasse 11, 6330 Cham, Switzerland

Geography of Cases in the Book

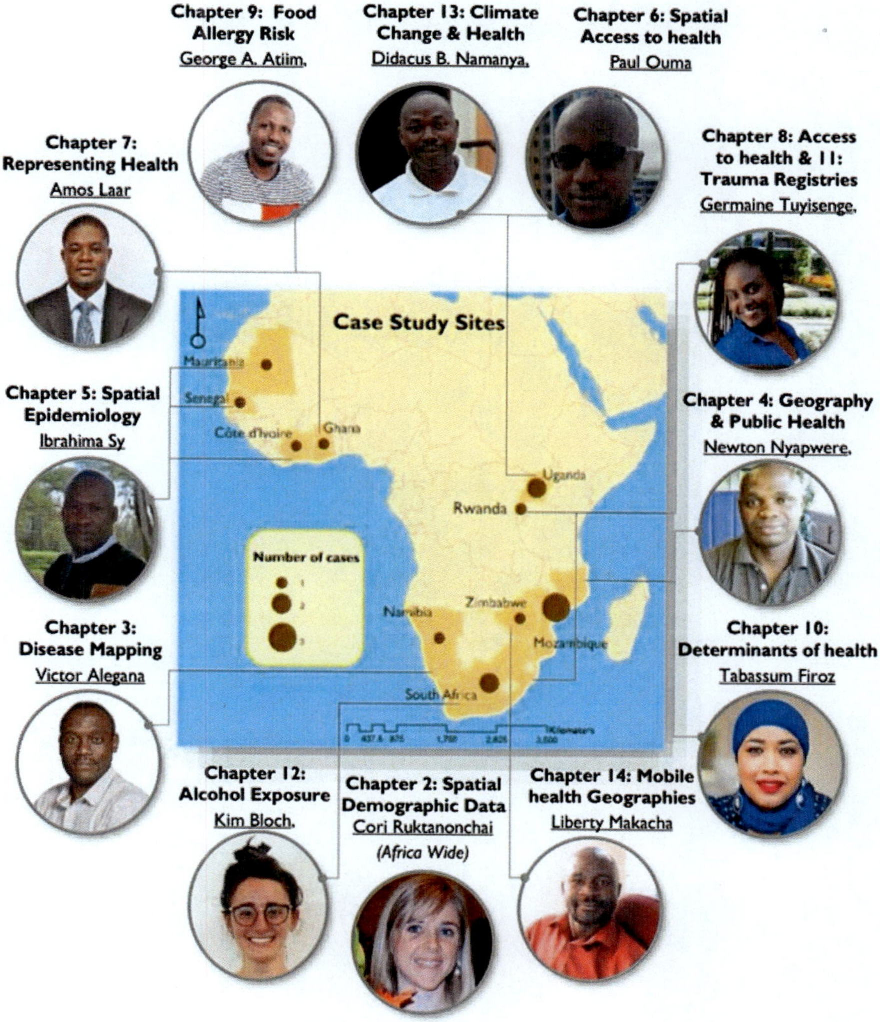

Fig. 1 Lead authors, themes and geography of cases in this book

About the Book

This book describes the characteristics of the practice of health geography in the African continent. Much of the story of geography of health in the African continent has been told through a burden of disease lens that normally presents a reactionary and aggregate perspective to health. Less is known concerning the different contextual forces that generate these patterns in disease. This edited volume will present 13 case studies and illustrations from different African countries to describe the various contemporary and traditional themes that have been known to characterize the discipline of health geography and how these look like in the African practice of the discipline.

Contents

Contributors

Victor Alegana is an Early Career Research Fellow at the KEMRI-Wellcome Trust Research Programme. He has a PhD in Geography and a Master's degree in Applied Geographic Information Systems and Remote Sensing both from the University of Southampton, UK. He has a keen interest in the use of routine health data and application of spatial–statistical data science to public health problems in low- and middle-income countries. This includes broad areas of population health particularly on spatial epidemiology (disease burden estimation), health care access, delivery of health interventions and monitoring sustainable development goals related to vulnerable populations.

George A. Atiim is a Postdoctoral Fellow with the United Nations University International Institute for Global Health (UNU-IIGH). At UNU-IIGH, he contributes to work on building conceptual clarity and approaches to promote gender mainstreaming in health programmes. His research broadly revolve around gender and health, the socio-cultural dimension of food allergies, food environments and the drivers of unhealthy diets and participatory method in sub-Saharan Africa.

Kim Bloch, MPH has over a decade of experience in public health research and programme management. Her special interests include urban upgrading as a violence prevention strategy as well as reproductive health. She was a researcher at the University of Cape Town's School of Public Health and Family Medicine from 2013–2018 and Programme Manager at VPUU from 2015 to 2017, which informed the chapter in this book. She now runs a social business creating economically empowering opportunities for organic, small-scale farmers through capacity building, sales and distribution of their produce across Cape Town, South Africa.

Tabassum Firoz is a Clinical Assistant Professor at the Yale School of Medicine in the Department of Medicine and Obstetrics, Gynecology and Reproductive Sciences. She is trained in Internal Medicine and Obstetric Medicine [medical disorders of pregnancy]. She obtained a Master's in Global Health Policy at the London School of Hygiene & Tropical Medicine and her research fellowship focused on global maternal health. She has published numerous papers, written book chapters and evidence-based policy briefs relating to medical disorders of pregnancy. She has been part of national guideline committees on the Hypertensive Disorders of Pregnancy. She serves on the Connecticut Maternal Mortality Review Committee and has been a technical advisor for the WHO Maternal Morbidity Working Group and WHO Near Miss Working Group.

Amos Laar has academic training in Nutrition, Public Health and Bioethics. In his independent scholarship, he draws on theoretical and methodological perspectives from the social sciences, bioethics and public health to understand how physical environment, social environment and structural forces affect health. His recent work in nutrition is centred on obesity and nutrition-related non-communicable diseases (NR-NCDs) prevention. He is the Principal Investigator of an IDRC-funded project, the MEALS4NCDs Project, which is providing **M**_easurement_ **E**_valuation,_ **A**_ccountab ility and_ **L**_eadership_ **S**_upport_ (**MEALS**) for NCDs prevention in Ghana. He was recently recognized in the Lancet Diabetes and Endocrinology for his efforts at combating NR-NCDs in Ghana: https://www.thelancet.com/journals/landia/article/PIIS2213-8587(19)30216-5/fulltext.

Liberty Makacha is Research Administrator at Midlands State University's Place Alert Labs. He is a PhD student at King's College, London, with a special interest in exploring the correlates between socio-environmental risk factors including ambient and indoor air pollution and the seasonal nature of adverse maternal outcomes in sub-Saharan Africa using a mixed methods approach. Liberty holds a Master of Science in Biostatistics and Epidemiology, a BSc (Hons) in Surveying and Geomatics both from the Midlands State University (Zimbabwe) as well as an Intermediate Bachelor of Accounting Science (BCompt-Intermediate) from the University of South Africa (UNISA).

Didacus B. Namanya is a skilled senior health geographer, with over 20 years experience working at the Ministry of Health Uganda. In this role, he is in charge of public health mapping, geospatial analysis and the focal person for climate change and human health. He holds MPH from Clarke International University, Kampala, Bachelor of Arts in Geography from Makerere University Kampala, Uganda (MuK), and a Postgraduate Diploma in Education (MuK). He is currently pursuing a PhD in Public Health at Uganda Martyrs University (UMU). Over the years, Mr. Namanya has been involved in several health and climate-related research projects at national and international levels resulting in numerous publications.

Newton Nyapwere is a GIS Analyst at the Midlands State University under the Place Alert Labs (PALs) Initiative within the Surveying and Geomatics Department. He has cross-cutting experience through applications of Geoinformation Science in health, agriculture and environmental management. He has a Master of Science in Applied Geoinformatics from Faculty ITC, University of Twente, and a Bachelor of Science (Hons) in Soil Science from the

University of Zimbabwe. He is currently pursuing a Master's degree in Information Systems Management at the Midlands State University. Newton has strong research interest in health geography, landscape epidemiology, and designing and developing GISystems for data acquisition, processing, manipulation, storage and retrieval and visualization of geoinformation.

Paul Ouma received training in Geomatics Engineering and Geospatial Information Systems from the Jomo Kenyatta University of Agriculture and Technology, Kenya. He later joined the Population Health group at the KEMRI-Wellcome Trust Research Programme as a research assistant, where he worked on mapping access to health facilities. He is currently a PhD student, where his work focuses on understanding the variation in spatial access to emergency hospitals for acute conditions in Africa. His work can be useful in providing important baseline information on where expansion for hospital and pre-hospital care should be focused, at both national and sub-national level. In addition, Paul is an active member of the Geospatial Information Management Society of Kenya.

Corrine Ruktanonchai received her Bachelor of Science degree in Ecology & Evolutionary Biology from the University of Tennessee in 2009 and completed a Master's in Public Health at the University of Florida in 2012. She completed her PhD in 2020 at the University of Southampton, where she utilized spatially explicit approaches to develop high-resolution maps of relevance to maternal and newborn health in East Africa. Currently, Cori works as a Research Associate with Novametrics Ltd., a research and knowledge management consultancy with expertise in designing, implementing, analysis and reporting on surveys, particularly within the context of maternal and newborn health.

Ibrahima Sy is currently Senior Lecturer in the Department of Geography at Cheikh Anta Diop University (Dakar, Senegal). He is also Associate Researcher at the Research for Development Department at the Centre for Ecological Monitoring (CSE) in Dakar, Senegal. Dr. Sy completed his PhD in Health Geography at the University of Strasbourg, France. Previously, he worked at the University of Strasbourg as Associate Professor. Dr. Ibrahima Sy coordinated several research projects tackling the challenges of water, environmental sanitation, spatial epidemiology, climate change and health in Mauritania, Senegal and Côte d'Ivoire in collaboration with several national institutions. The recent research project that he coordinated at the CSE is ACASIS. Ibrahima's area of interest is in Environmental Sanitation, Spatial Epidemiology, climate change and Urban Health.

Germaine Tuyisenge is a PhD candidate in Health Geography at Simon Fraser University. She also holds an MA in Health Geography from Western University, Canada. Germaine is also a Fellow of the Queen Elizabeth Diamond Jubilee Scholarship (Kenya, 2015). Her research focuses on micro-level dynamics that facilitate women's access to health services in resource-limited settings. Prior to undertaking graduate studies, Germaine worked with government and non-government institutions in Rwanda (her home country), Kenya, the Netherlands and Canada. This work has focused on community health promotion, sexual and reproductive health and women's access to health services. Germaine is interested in the spatial and place-based dimensions of health and health care. Her research focuses on the social determinants of health and using community-based practices to eliminate health inequities experienced by marginalized groups.

About the Editor

Prestige Tatenda Makanga is an applied geoinformation scientist with a primary interest in global health. He is Chairman and Senior Lecturer in the Surveying and Geomatics Department at the Midlands State University (MSU) in Gweru, Zimbabwe. He leads the Place Alert Labs (PALs—https://ww5.msu.ac.zw/pals) initiative at MSU. The PALs programme is a mixed methods health geography initiative that explores how places shape local health experiences and health outcomes. Much of the evidence generated from the work at PALs is translated into decision aids and tools that are used to target health interventions.

Chapter 1
Introducing Health Geography

Prestige Tatenda Makanga

Abstract This introductory section summarizes the key characteristics of health and medical geography. It also illustrates how these characteristics have shaped what's currently known concerning the practice health geography in Africa. The chapter ends by briefly describing the different chapters of the book and how these illustrate the diverse character of the practice of health geography in Africa.

Keywords Health geography · Medical geography · Health and space · Health and place · Geo-enabled health decisions

Health and Medical Geography

Health geography is a sub-discipline of human geography, which offers a holistic view on health by linking health and disease outcomes to the socio-cultural and physical environment, and the places that generate them (Dummer 2008).

Historically, the terms 'medical geography' and 'geography of disease' were used interchangeably to describe the discipline (May 1950), which had much influence from biomedical models of health, and was driven largely by the disease ecology framework (Meade 2012). The work of medical geographers has therefore naturally been more empirical and quantitative in nature, and has been more popular with funders of health research for the same reasons that interventions of a biomedical nature attract more funding compared to non-clinical ones (Kearns and Moon 2002). The work of medical geographers thus features prominently in medical journals, which have greater impact factors than the geography journals, maintaining the medical geographers' visible presence and relevance, especially to a clinical and epidemiological audience, while remaining conspicuous through an absence from geography literature.

P. T. Makanga (✉)
Place Alert Labs (PALs), Department of Surveying and Geomatics, Midlands State University, Gweru, Zimbabwe
e-mail: makangap@staff.msu.ac.zw

P. T. Makanga (ed.), *Practicing Health Geography*, Global Perspectives on Health Geography, https://doi.org/10.1007/978-3-030-63471-1_1

It is close to three decades since the landmark call for a reformed medical geography was made by Kearns (1993). A major prompt for this call was a claim that medical geographers had a too simplistic interpretation of space as a container of health and illness and had neglected the role that context plays in generating good and bad health outcomes (Kearns and Joseph 1993). Space is defined as 'the dimension within which matter is located or a grid within which substantive items are contained' (Agnew 2011). In medical geography, space has been the organizing principle for health data, mainly through using it as the basis for describing prevalence rates and patterns for disease within distinct spaces (Dorn 1994), an approach also commonly used in global disease burden studies. In other words, spatial (or space based) data is used to describe the composition of disease within distinct spaces (Macintyre et al. 2002). It has been argued that this use of space as a container for disease count results in aggregate measures of disease trends, and aggregate analysis is essentially 'incapable of distinguishing the contextual (the difference a place makes) from the compositional (what is in a place)' (Jones and Moon 1993).

Place is a more subjective and less intuitive term that speaks specifically to the contextual rather than compositional matters that make up spaces (Macintyre et al. 2002). It encompasses the social and cultural characteristics that influence health and health care delivery (Andrews et al. 2012). The need for mechanisms to traverse through a place and acquire knowledge about the interactions between health and the contextual forces that generate it was one of the things that prompted the need to define a new health geography research agenda. The new health geography which values mechanisms for exploring health through the lens of place goes beyond the geometric construction of space to understand the contextual forces behind good or bad health outcomes (Kearns and Joseph 1993). With place being the central guiding rod for health geography research and space for medical geography, a different set of methods are preferred in each of the two sub-disciplines. Health geography mainly uses methods from human geography; for example narratives and storytelling (Kearns 1997) or focus group discussions (Vuksan et al. 2012). However, studies that are aligned with medical geography mainly utilize methods rooted in the natural sciences and spatial epidemiology like Geostatistics (Shoff et al. 2012) and mathematics (Berke 2004). These differences in methods have somewhat made the chasm between the two disciplines more apparent.

Three decades on, it remains unclear whether or not there has been broad consensus on what should constitute a reformed geographies of health between the health and medical geographers. What is apparent though, is that both streams of work are still happening, and that both offer useful perspectives into the geographic inquiry of health.

Purpose and Overview of Book

This edited volume will present 13 case studies and illustrations from different African countries, to describe the various contemporary and traditional themes that have been known to characterize the discipline of health geography and how these look like in the African practice of the discipline. Many of the challenges and opportunities that are somewhat unique to the African contexts represented in the book are used to illustrate some the issues that are not necessarily considered as part of the mainstream practice of health geography.

This is the first time that a single volume will be dedicated to the practice of health geography in Africa. Each chapter in the book represents a unique case (or cases) describing both how different African contexts are shaping the practice of health geography on the continent, and the different contextual factors and their effects on health in different places in Africa. Some of the chapters were written by authors who do not necessarily self-identify as health geographers but nonetheless have adopted a geographical perspective to different responses to population health needs. This uniquely illustrates the extending reach of the discipline and in many ways illustrates how it is being shaped by multiple disciplines outside geography.

The book has also started by giving an overview of the known debates within health and medical geography discipline, but goes further to situate these within the geographic inquiry into health on the African continent. This was done in a manner that challenges this imaginary dichotomy and chasm between medical and health geography that these debates seem to have created. The unique examples from Africa will illustrate the value of a unified perspective to the geographies of health.

The book also illustrates the unique value that geographic thought has in operationalizing some of the core values and priorities of global health using examples from African cases. These include:

(a) How place and space based geographical perspectives both play an important role in linking determinants of health to health outcomes.
(b) How geography is useful for implementing the strategies within the new sustainable development goals, e.g. how it enables us to 'draw on contributions from indigenous peoples, civil society, the private sector and other stakeholders, in line with national circumstances, policies and priorities'.
(c) How geographic methods allow for implementing the call for greater measurement of disaggregate subnational trends in life-course health outcomes and associated determinants as emphasized in the new sustainable development goals.

Chapter Summary

This **introductory** section has summarized the key characteristics of health and medical geography and how these link up with perspectives in global health. The following chapters present cases that illustrates how these characteristics have shaped what's currently known concerning the practice health geography in Africa.

Part I "**Health and Space**" consists of five cases illustrating different uses of space-based analyses in health geography. Mostly, the chapters showcase quantitative approaches to understanding the geographies of health. In this vein, Chap. 2 illustrates novel methods for creating high resolution demographic GIS datasets that are fundamental for much health geography research and intervention. Specific methods that were explored include the use of satellite imagery and machine learning, as well as cellphone data to derive fundamental geo-demographic data and track movement of populations.

The role of the Geoinformation sciences and geography in describing the malaria disease burden in Namibia is explored in Chap. 3. The authors illustrate how geographically coded health records are being incorporated into disease surveillance and how this translates into maps of disease burden.

Chapter 4 provides a number of cases that illustrate the value that spatial thinking brings to epidemiological analyses linking health outcomes to determinants. While the field of spatial epidemiology is well established in many regions of the world, the full value of explicit spatial analyses as part of epidemiologic inquiry is yet to be fully demonstrated in Africa. This chapter provides empirical cases that differentiate spatial epidemiology from traditional epidemiology illustrating the value that geography brings to the process.

'Precision public health' is a new buzz word that attempts to bring the gains in precision medicine into public health. Geography has potential to facilitate more geographically precise responses to public health problems. This is particularly important for places with high disease burden but very little in the way of resources to respond to the problem. Chapter 5 illustrates this intersection between geography and precision public health through a case that illustrates the value that geography brings to precision public health in regions of Southern Mozambique.

Chapter 6 describes the use of geospatial techniques in evaluating spatial accessibility to health services. While modelling spatial access to health care is one of the main applications of GIS in health, the vulnerability of transport infrastructure and the impact of seasonal elements like precipitation and floods have rendered many of the conventional access models inadequate for a typical African setting. Furthermore, the data challenges in many African settings have resulted in the development of new techniques for filling these data gaps and modelling access that are suitable and can be scaled across Africa. This chapter illustrates the novel methods that cater for the unique geographies in Africa.

Part II "**Health and Place**" consists of four cases that illustrate different uses of place based analyses in health geography. Mostly, the chapters will showcase qualitative and potential for mixed approaches to understanding the geographies of health.

Chapter 7 is on representing health and it challenges the dominance of the disease perspective as the manner for imagining health. The case study provides and Afrocentric perspective of how health is more than just the absence of disease and exposes, how issues of social location, spiritual connections and other socio-cultural determinants contribute to shaping individual and populations' experience and understanding of health.

While Chap. 6 addresses the geographical challenges in accessing care, Chap. 8 uses a case in Rwanda to explore community dynamics pertaining to access of maternal health services. Community health workers, who are volunteers elected within their own communities, provide maternal health services to women and collaborate with other community members to link communities to the formal health care system. The mostly nonphysical geographical barriers to accessing care are explored.

Chapter 9 explores the interactions between place and health through the use of Photovoice, a participatory research exercise for documenting people's health experiences by having them take photographs that represent these experiences. This is followed up by semi structured in depth interviews to elucidate how these photographs symbolize lived health experiences, highlighting the multiple spaces and places of health risk exposure.

The new paradigm of global health offers the promise for improving health by addressing the social determinants of health that generate the visible patterns in health outcomes. One of the main challenges with this paradigm though concerns how to know and measure these determinants of health. Chapter 10 provides the background to the social determinants of health and illustrates how this is a geographical problem at multiple scales. The case study presented in this chapter describes how to know what determinants are related to maternal health in regions of Southern Mozambique.

Part III "**Geo-Enabling Health Decisions**" illustrates real examples of how geography is being incorporated into decision processes on the continent. These include how evidence is being used to alter policy priorities and plan for health-related interventions at both the population and individual level.

Chapter 11 describes the value of incorporating geography into health surveillance. Using examples from trauma surveillance in Cape Town, South Africa, the chapter illustrates the mass transition that is required within health systems for them to adapt to the spatial turns in health, what this looks like operationally and what the short and long-term benefits of doing this are.

Chapter 12 focuses on alcohol as a key determinant of injury related to interpersonal violence in Cape Town, South Africa and how different geographical measures of access to illegal liquor outlets are used to develop policy recommendations aimed at reducing alcohol related violence. In addition to translating evidence from clinical studies into policy positions, examples are given that illustrate how evidence from spatial and place based geographical methods are being translated into policy for improving health.

Chapter 13 describes a case of how health geography is being used to plan for population level health interventions in regions of Uganda. In particular the author emphasizes the intersection between climate change and public health planning, and how geographic thought could help to take targeted action to mitigate the potential negative effect of the changing climate on health.

Chapter 14 concludes the book with two fascinating cases of geographically enabled mobile health tools. The ubiquity of mobile telephony on the continent and the increasing availability of location-based services on mobile phones is changing how people are experiencing health. Mobile phones are linking people in isolated regions of Africa to information that enables them to make health-related decisions. This chapter illustrates the potential of geography in aiding mobile health through examples that translated evidence from Chaps. 5 and 11 into tools to improve access to maternal care and address community specific determinants of maternal ill-health. These tools were piloted in Zimbabwe and Mozambique respectively and help to uniquely illustrate the power of a unified geographies of health that embraces both place and space as part of geographic inquiry into health.

References

Agnew, J. (2011). Space and place. *The Sage Handbook of Geographical Knowledge*, 316–330.

Berke, O. (2004). Exploratory disease mapping: Kriging the spatial risk function from regional count data. *International Journal of Health Geographics, 3*(1), 18.

Dorn, M. (1994). Social theory, body politics, and medical geography: Extending Kearns's invitation. *Professional Geographer, 46*(1), 106.

Dummer, T. J. B. (2008). Health geography: Supporting public health policy and planning. *Canadian Medical Association Journal, 178*(9), 1177. https://doi.org/10.1503/cmaj.071783.

Evans, A. G. J. J. et al. (2012). Arguments in Health Geography: On Sub-Disciplinary Progress, Observation, Translation. *Geography Compass, 6*(6): 351–383.

Jones, K., & Moon, G. (1993). Medical geography: Taking space seriously. *Progress in Human Geography, 17*(4), 515–524. https://doi.org/10.1177/030913259301700405

Kearns, R. A. (1993). Place and Health: Towards a Reformed Medical Geography*. *The Professional Geographer, 45*(2), 139–147. https://doi.org/10.1111/j.0033-0124.1993.00139.x

Kearns, R. A. (1997). Narrative and metaphor in health geographies. *Progress in Human Geography, 21*(2), 269–277. https://doi.org/10.1191/030913297672099067

Kearns, R. A., & Joseph A. E. (1993). Space in its place: Developing the link in medical geography. *Social Science & Medicine, 37*(6), 711–717.

Kearns, R., & Moon, G. (2002). From medical to health geography: Novelty, place and theory after a decade of change. *Progress in Human Geography, 26*(5), 605–625. https://doi.org/10.1191/0309132502ph389oa

Macintyre, S., A. Ellaway, et al. (2002). Place effects on health: how can we conceptualise, operationalise and measure them? *Social Science & Medicine, 55*(1), 125–139.

May, J. M. (1950). Medical geography: its methods and objectives. *Geographical Review, 40*(1), 9–41. JSTOR. https://doi.org/10.2307/210990

Meade, M. (2012). The geography of life and death: deeper, broader, and much more complex. *Annals of the Association of American Geographers, 102*(5), 1219–1227. https://doi.org/10.1080/00045608.2012.687351

Yang, S. C. T. -C., et al. (2012). What has geography got to do with it? Using GWR to explore place-specific associations with prenatal care utilization. *GeoJournal, 77*(3), 331–341.

Williams, V. M. A. M., et al. (2012). The workplace perspective on supporting family caregivers at end of life: evaluating a new Canadian social program. *Community, Work & Family, 15*(3), 317–333.

Part I
Health and Space

Chapter 2
Generating Spatial Demographic Data for Health in Africa

Corrine Ruktanonchai, Andrew Tatem, Victor Alegana, and Zoe Matthews

Abstract Spatial data at various policy relevant administrative units are fundamental to designing, implementing and evaluating health programs and interventions throughout low- and middle-income countries. Many countries in sub-Saharan Africa, however, face unique challenges in developing and utilizing spatial data, potentially hampering the implementation of health Geographic Information Systems (GIS) infrastructure and research. In this chapter, we explore the use of spatial demographic data within the African context by firstly outlining the utilization of GIS and availability of spatial demographic data for Africa, followed by the evolution of methods and current approaches to estimating demographic data using ancillary data sources, as well as challenges unique to spatial demographic data. We then present five case studies highlighting innovative methods of production and use of freely available spatial demographic data to address policy needs and health challenges. Finally, we highlight emerging future directions in the production and improvement of spatial demographic data, followed by recommendations for enhancing spatial data infrastructures for health applications within low- and middle-income country settings.

Keywords Spatial data · GIS · Health geography · Spatial data infrastructures · Disease mapping

C. Ruktanonchai (✉)
Geography and Environment, University of Southampton, Southampton, UK

A. Tatem
Population Health Unit, Kenya Medical Research Institute - Wellcome Trust Research, Nairobi, Kenya

V. Alegana
Population Health Unit, Kenya Medical Research Institute - Wellcome Trust Research, Nairobi, Kenya

Geography and Environment, University of Southampton, Southampton, UK

Z. Matthews
Social Statistics and Demography, University of Southampton, Southampton, UK

P. T. Makanga (ed.), *Practicing Health Geography*, Global Perspectives on Health Geography, https://doi.org/10.1007/978-3-030-63471-1_2

Introduction

With recent advancements of more computationally efficient resources and user-friendly applications, the past decade has seen an exponential rise in applications of geographical information systems (GIS) towards addressing public health research questions (Banerjee 2016). Adoption and uptake of GIS techniques within public health has generally been slower within Africa, however, due to a number of obstacles such as infrastructure and cost constraints, lack of donor investment involving capacity building and local engagement, and lack of sustainable research funding moving past the donor-led phase of research projects (Tanser and le Sueur 2002). Despite this, GIS tools offer great potential in addressing some of the most significant public health threats facing the African continent, such as malaria, tuberculosis and HIV/AIDS, which are often characterized by having strong environmental components. Over the coming decades, data visualization through disease mapping and spatial analysis can play an increasingly significant role in ensuring that preventable deaths are avoided and disease burden is lessened amongst the most vulnerable populations (Molla et al. 2017; Tanser and le Sueur 2002). This necessitates, however, that challenges are adequately addressed and recommendations made in ensuring high quality spatial data that can be collected even within resource-poor settings.

In this chapter, we address current and emerging innovations in freely available spatial demographic data production that can be used in GIS applications throughout Africa. Firstly, we outline the utilization of GIS techniques and availability of spatial demographic data in Africa, followed by the evolution of methods and current approaches used to estimate these data, where other reliable sources might be lacking. We then present five case studies highlighting innovative applications in producing freely available spatial demographic data, including subnational age and gender composition, high-resolution live births and pregnancies, and quantifying human mobility derived from mobile phone data. Lastly, we discuss emerging future directions in improving the production of spatial demographic data, and its applications to low-resource settings and policy-making.

GIS Utilization and the Availability of Spatial Demographic Data in Africa

GIS Utilization Within Africa

With an increasing number of health researchers adopting GIS techniques, potential continues to grow in exploring the spatial variation of disease and its relationship to the environment, population and health systems. The application of these methods can be limited, however, by infrastructure and cost constraints, as well as specialist knowledge capacity. While GIS can help to fill these gaps, the technology must be aimed at health problems and priorities relevant to the region. Towards this, Tanser and le Sueur (2002) performed a systematic review of the health literature within African settings at the turn of the millennium, examining GIS research amongst top public health threats within Africa, such as HIV/AIDS, malaria and tuberculosis. Overall, they found that GIS applications have been widely applied within Africa to diseases with strong environmental components, such as malaria elimination settings, but fewer studies have utilized GIS to study non-vector borne diseases such as tuberculosis and HIV/AIDS. Despite the ability to address research questions central to African health issues, they note that the use of GIS techniques remained underutilized, and suggested a need for greater capacity building efforts to train GIS experts within-countries (Tanser and le Sueur 2002).

Yet with global trends favouring more powerful computational techniques and hardware, greater availability of user-friendly software, and increasing reliability of demographic data collection and vital health registrations, the use of GIS continues to hold promise for addressing health research priorities and policy planning within Africa (Tanser and le Sueur 2002). More recently, Ebener and colleagues performed a rapid literature review over a decade later to examine utilization of GIS techniques within the field of maternal health (Ebener et al. 2015). Overall, they note that despite the increasing interest and promise in the use of GIS techniques to address health problems, methods remained largely under-utilized within the field of maternal and newborn health in low- and middle-income countries. Specifically, they report three primary emerging areas of research within the field: (1) thematic mapping (i.e. choropleth maps conveying information on a topic or theme), (2) spatial analysis (i.e. creation of new information using spatial data) and (3) spatial modelling (i.e. statistical or mathematical models simulating a spatial process or phenomena). Despite the emergence of these broad research areas, they recommend better communication and coordination amongst institutions to build GIS capacity and expertise, including amongst in-country health sectors and non-academic institutions (Ebener et al. 2015).

Availability of Spatial Demographic Data

However, to facilitate greater application and uptake of GIS methods in addressing health research questions specific to Africa, reliable, high-resolution and contemporary spatial data is necessary. Over recent decades, the use of spatially explicit demographic data, particularly quantifying human population distributions, has seen broad applications across spatial epidemiology, urban planning, accessibility modelling, disaster management, resource allocation and more (Tatem et al. 2007). Towards this, several sources of spatially explicit data exist to inform demographics such as population counts, age and gender composition, urbanization, etc., including *census data, census microdata* and *survey data.*

Census data consist of the most detailed information regarding spatial demographic data within a country, but are typically carried out around once every 10 years (Tatem et al. 2012). Further, while high-income countries have extensive resources available to conduct census surveys, low-income countries can suffer from constraints in resources, lack of knowledgeable experts, and competing development priorities, with the most recent mapping and census efforts for some countries occurring 30–40 years ago (Tanser and le Sueur 2002; Tatem et al. 2007).

In addition, these data are often aggregated to coarse administrative levels, and can be proprietary and difficult to obtain. To supplement these challenges as well as fill in temporal gaps, *census microdata* are comprised of large, representative subsamples of households surveyed during a census generally stored in data repositories like the International Public Use Microdata Series (https://international.ipums.org/international/), making these data easier to obtain than full censuses. These data may be available at smaller spatial units than typical surveys, with similar levels of spatial demographic data such as age, gender, population, etc. (Tatem et al. 2012). Lastly, *survey data* collected through international survey programmes such as the Demographic and Health Surveys (DHS), Multiple Indicator Clustery Survey (MICS) and AIDS Indicator Survey (AIDS) are collected on regular intervals and standardized across countries to allow for comparison across borders. Not only are these data rich in demographic and health information on local populations, but they are becoming increasingly spatially detailed, as represented by geo-referencing of clusters among DHS surveys. Further, these data are freely available to researchers across the world, and therefore represent a promising avenue for obtaining spatial demographic data, particularly amongst developing countries (Fig. 2.1).

However, in the absence of contemporary sources of spatial data, in addition to decentralized, uncoordinated efforts which can often lead to conflicting, unreliable data sources, the production of spatial demographic data in resource-poor settings can oftentimes rely upon augmentation with ancillary, satellite-derived data sources. These data sources may include human settlements, distance to nearest urban centre, night-time lights and more, and are often employed in geostatistical models to predict distributions of human populations, including age and gender composition, births, pregnancies, etc. Mapping efforts utilizing these data sources have grown over the past few decades, evolving from simple areal weight redistribution methods to more sophisticated machine-learning approaches requiring computationally intensive time and resources. In the following sections, we discuss the evolution of efforts to map demographic data, as well as current approaches in quantifying human population distributions at a high spatial resolution.

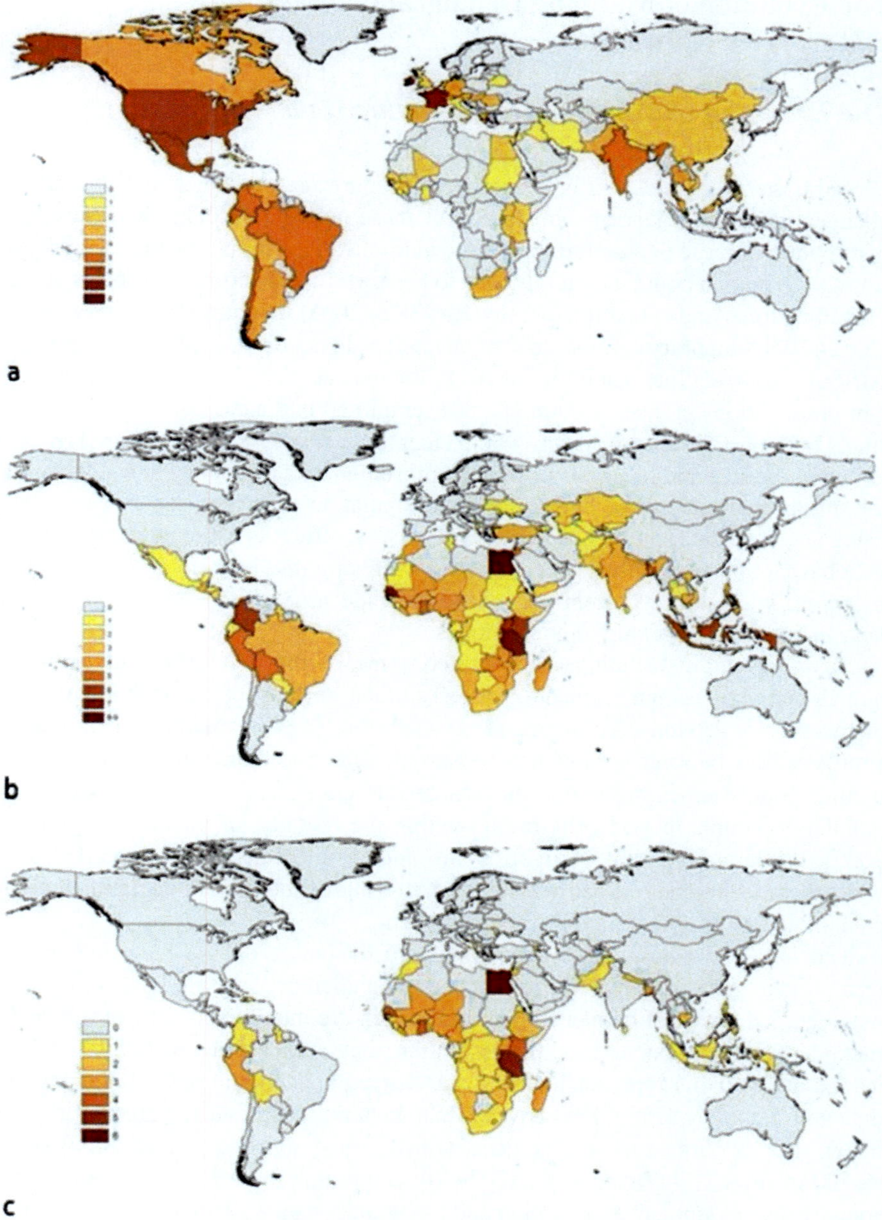

Fig. 2.1 Availability of demographic datasets informing subnational population attributes, among (**a**) census microdata, (**b**) national household survey programmes and (**c**) national household survey programmes with corresponding GPS information. (Adapted from Tatem et al. 2012)

The Evolution of Mapping Demographic Data and Current Approaches

The Evolution of Mapping Demographic Data

Global populations are rapidly changing, with increasing urbanization, as well as changes in age and gender composition of these populations. The dynamics and distribution of these populations have crucial implications for public health applications, with human population totals used to generate disease burden estimates at the global, national and subnational scale (Hay et al. 2005). Oftentimes, however, reliable and contemporary census and settlement data do not exist, particularly amongst African countries (Tatem and Linard 2011; Tatem et al. 2007). This can have important ramifications in disease outbreak, as exemplified in the case of the 2013–2016 Ebola outbreak in Western Africa, where emergency response was hampered by not knowing the size and locations of population settlements (Varshney et al. 2015). In the absence of such data, high-resolution population estimates can be produced using ancillary data sources available. Historically, efforts to estimate demographic data have employed methods such as *areal weighting and land-cover-based redistribution,* while more current approaches include *dasymetric redistribution* and *bottom-up population mapping* techniques.

Areal weighting techniques constituted some of the first efforts to quantify human populations on a gridded (or raster) surface, such as the Gridded Population of the World, version 2 (Deichmann et al. 2001). These techniques consist of a simple overlay between a gridded raster surface and an administrative polygon containing population information (such as census units), and evenly redistributing population counts to grid cells falling within the polygon of interest (Hay et al. 2005). While conceptually straightforward and easy to calculate within GIS software, these techniques critically assume a uniform distribution of population within an administrative unit, which is often unrealistic.

The *land-cover-based redistribution* approach builds on the areal weighting technique to generate population estimates using ancillary satellite imagery data to map detailed layers of human settlements, which are then linked with census estimates to generate accurate and high-resolution population estimates. Tatem and colleagues utilized this approach in the East Africa region, combining satellite imagery data with remotely sensed land-cover data to generate reproducible, adaptable estimates of populations across large geographic areas at high spatial resolutions such as 100 m (Fig. 2.2) (Tatem et al. 2007). By combining these freely available data sources such as land cover, elevation and topography data, they demonstrated that it was possible to generate accurate estimates of human settlements and population estimates at a higher spatial resolution than any previous work. While this approach was widely scalable and flexible across low- and middle-income countries, it relied heavily on input data sources such as land cover and climate. In light of this, this approach has generally been replaced by an alternative, machine learning algorithm

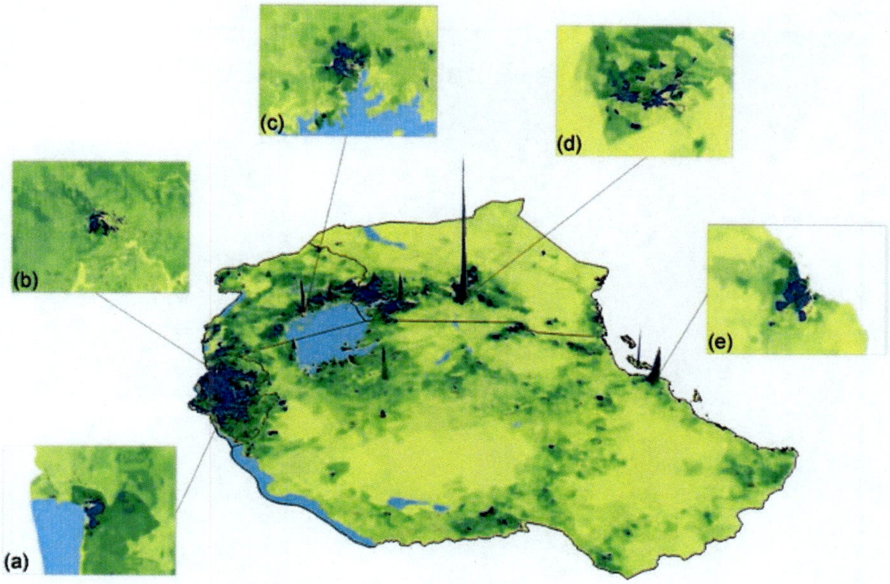

Fig. 2.2 Population density of the East Africa region (Burundi, Kenya, Rwanda, Uganda, Tanzania), with selected inset cities, (**a**) Bujumbura, (**b**) Kigali, (**c**) Kampala, (**d**) Nairobi, and (**e**) Dar Es Salaam, using a land-cover-based redistribution approach. (Adapted from Tatem et al. 2007)

such as the random forest approach, which incorporates data from a wider range of ancillary data, including distance to health facilities, schools, road networks and more.

Current Approaches to Mapping Demographic Data

Stevens et al. outline a random forest mapping technique used to disaggregate census data and generate high-resolution population estimates for Kenya, Vietnam and Cambodia (Fig. 2.3) (Stevens et al. 2015). This approach utilizes a semi-automated modelling approach, incorporating continuous global datasets, such as night-time lights, land cover and topography, as well as discrete covariates, such as distance to the nearest urban area. By combining these data with census data, this approach employs *dasymetric redistribution*, where population counts are 'intelligently' redistributed from a coarser spatial resolution (such as census enumeration areas) to a finer spatial resolution, while preserving the total counts at the original input source (Mennis 2003). However, instead of redistributing these population counts evenly across the finer spatial area of interest, a regression tree model (or random forest) can be used to 'unequally' distribute counts, or weight their distribution (Fig. 2.4) (Stevens et al. 2015). This weighted layer contains information from a range of ancillary sources, as outlined above, and reflects underlying mechanisms

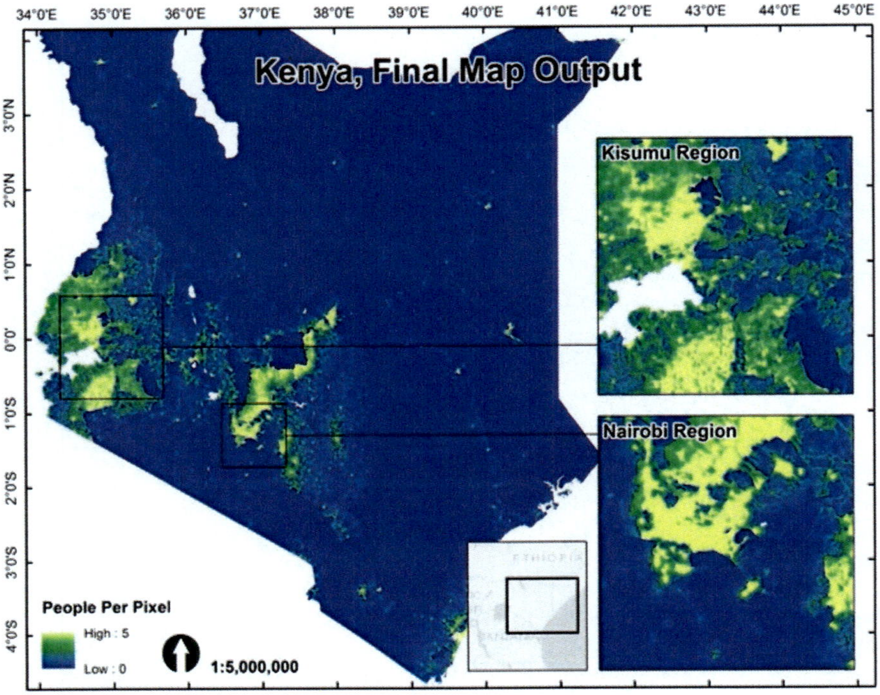

Fig. 2.3 High-resolution populations counts redistributed using Kenya 1999 census data and ancillary data sources (freely available at worldpop.org.uk). (Adapted from Stevens et al. 2015)

contributing to unequal population distributions. This approach is robust, semi-automated and scalable, and shows improvement over more traditional mapping approaches, such as previously outlined land-cover-based techniques (Stevens et al. 2015).

Where census data are unreliable or out of date, *'bottom-up' population mapping* techniques have been employed, as outlined by Wardrop et al. (2018). Similar to the previous land-cover-based and dasymetric redistribution techniques as outlined above, the goal of bottom-up population mapping is to generate high-resolution population estimates using a range of robust, and ideally freely available, data sources. In contrast to the above approaches (which can be considered 'top-down' population mapping), bottom-up population mapping relies on micro-census data, which collects information within smaller regions than census enumeration areas, collected at more frequent and rapid intervals (Fig. 2.4) (Wardrop et al. 2018). Using statistical models to account for residual spatial autocorrelation, these micro-census population counts are linked to spatial covariates to predict population counts in un-sampled locations. Few studies have validated outputs or accuracy of these methods over other techniques such as top-down approaches, however, suggesting an emerging area for future research. Regardless, bottom-up approaches offer potential for predicting population counties in countries where census data are

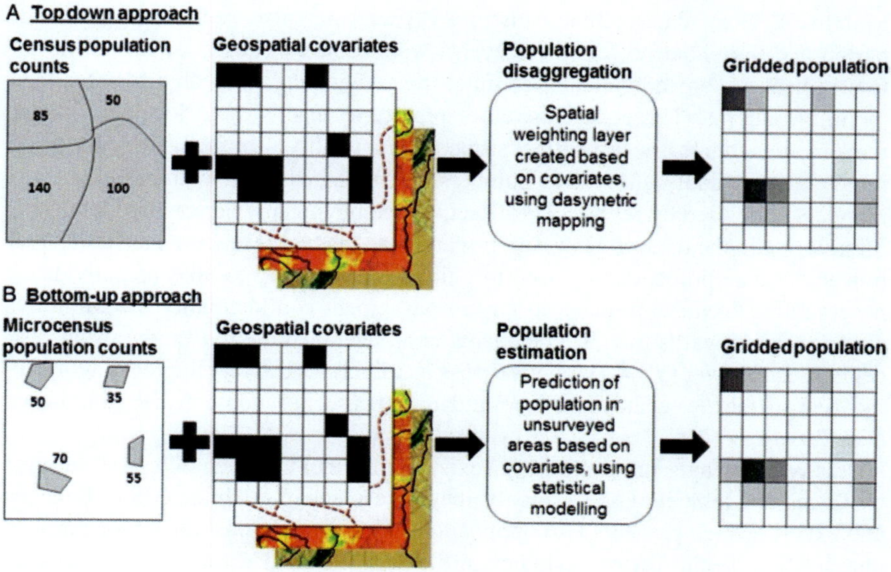

Fig. 2.4 Top-down (**a**) versus bottom-up (**b**) approaches to mapping gridded populations. (Adapted from Wardrop et al. 2018)

unreliable or sparse, offering a robust method to predict into un-sampled locations with associated model errors and uncertainty.

Challenges in Spatial Data Production

Maintaining participant confidentiality, particularly when data contain sensitive health or identifying information, is a crucial consideration for any database. As spatial databases generally contain some sort of locational information, confidentiality is an important challenge, particularly amongst datasets containing highly localized movements such as call data records (CDRs) or human movement patterns. In some cases, removing patient identifiers such as name, address, and date of birth, is a reasonable step to protecting confidentiality—however, in the case of locational coordinates or other spatially identifiable situations, this may not be an option. Approaches exist, however, to overcome these challenges and effectively anonymize data while maintaining relevant spatial information, such as those used by Ruktanonchai and colleagues to anonymize CDR data (Ruktanonchai et al. 2016).

Working with the primary mobile phone provider across Namibia (Mobile Telecommunications Limited) representing nearly 85% of the mobile phone market, Ruktanonchai et al. obtained an anonymized dataset consisting of 1.2 million unique SIM cards in Namibia representing over 9 billion communications from October 2010 to September 2011 (further details outlined in 'Quantifying Human

Mobility'). These data include calls or SMS text messages sent during the study period and the phone towers these events were routed through. By observing the towers used by an individual over time, the researchers were able to infer their mobility patterns. These data therefore represent sensitive and localized human movement patterns, necessitating strict confidentiality. To facilitate this, MTC removed individual SIM phone numbers and replaced them with random study numbers to be used by researchers, effectively removing any personally identifying information in the dataset. However, research has suggested that even as few as four movement data points can be used to uniquely identify up to 90% of individuals, necessitating further protocols to ensure participant confidentiality (de Montjoye et al. 2013). Towards this, Ruktanonchai et al. averaged the movement of all residents at each tower ($n = 402$ towers across Namibia), effectively aggregating movement data to the tower catchment level and further ensuring anonymity (Ruktanonchai et al. 2016).

Lastly, challenges unique to Africa exist in uptake of GIS technology, as outlined by Tanser and le Sueur (2002). Firstly, they note a lack of qualified GIS technicians and experts which prevents GIS applications from moving past the donor-initiated phase. Many health research studies in the area are funded through international donor agencies, and are often led by non-African scientists and principal investigators with sparse local engagement and contribution. To facilitate sustainable GIS applications to address local health problems, capacity training and support must be prioritized such that local scientists who understand both cultural and health systems factors can prioritize the research agenda. Secondly, they note a lack of suitable spatial datasets in addressing health and geographical research, which suffer from decentralized and uncoordinated efforts, in addition to time and cost constraints. Tanser and le Sueur argue that cross-sector projects may help to address these challenges, creating a more unified and systematic approach to creating large spatial databases. They further argue cost-effective methods of creating these datasets include establishing large-scale sentinel demographic and health surveillance datasets. Molla et al. make similar recommendations for spatial data used in addressing maternal and newborn health questions, suggesting universal datasets should be created with common data features such as geo-coding, health facility lists and more (Molla et al. 2017).

However, in the absence or interim of strengthening sustainable capacity building efforts, it is vital to ensure that high-resolution, contemporary and reliable spatial demographic data being produced are made available to researchers across the globe. Towards this, the WorldPop project (www.worldpop.org) based at the University of Southampton works to ensure that every person is accounted for in the decision making and resource allocation process. Fundamental to this vision is making both spatial demographic data and methods open, freely available and transparent. The datasets available through the WorldPop project have been used not only by academic researchers across low- and middle-income countries, but also policy makers, statistical agencies and international development agencies across the world. In the following section, we highlight five case studies outlining the production and application of datasets generated through the WorldPop project, demonstrating how these innovative datasets can be used for cost-effective decision making and resource allocation efforts, particularly in the absence of other reliable data sources or while capacity building and training efforts are ongoing.

Applications in Producing Spatial Demographic Data

Despite the growing availability of spatial data, the problem of how to generate data for un-sampled locations persists. While national-level household surveys such as DHS surveys are increasingly employing geo-referencing techniques, data are often presented at the national or regional level, and may mask important inequalities at higher spatial resolutions (Bhutta and Reddy 2012). The production of spatial demographic datasets has therefore become an increasingly recognized tool to fill in these spatial gaps by generating high-resolution surfaces of environmental covariates, population estimates, health outcomes and more (Ebener et al. 2015). Here, we present five case studies employing innovative methods to generate high-resolution spatial demographic data depicting subnational age/gender composition and high-resolution births and pregnancies, as well as quantifying human movement patterns through mobile phone datasets.

Mapping Spatial Demographics

Health metrics and intervention planning often aim to target specific populations, such as family planning interventions targeted towards women of childbearing age. In addition to knowing where populations are, it is therefore also vital to know who comprises these populations. Historically, however, studies mapping indicators and health risks have used statistics at the national level, and do not account for localized demographic variation and may contribute to hidden inequalities amongst the most vulnerable populations. Demographics such as age, gender and estimates of births and pregnancies have important monitoring and surveillance applications, such as malaria surveillance, vaccination coverage planning and family planning interventions, amongst many others. Capturing subnational variation in these populations is therefore crucial to ensure effective resource allocation. Here we outline several current approaches in quantifying these populations at subnational and high-resolution spatial scales.

Age and Gender Composition

To quantify subnational variation in age and gender composition, Tatem et al. constructed a detailed and contemporary population dataset for the African continent incorporating information on subnational age and gender structure (Tatem et al.

2013). To do this, they compiled and collated estimates for age and gender from a range of data sources representing more than 20,000 subnational administrative units, primarily via census data, household-level microdata collected via census surveys, and census microdata. Where contemporary census data were not available, however, national household survey data were used, such as those collected through the DHS, Malaria and AIDS Indicator Surveys, and Multiple Indicator Cluster Surveys. These subnational estimates and proportions were then linked to spatial information showing the corresponding administrative boundary, where existing population datasets as described above could be adjusted to display high-resolution information on 2010 age and gender structure for all continental Africa countries, plus Madagascar (Fig. 2.5) (Tatem et al. 2013). Additionally, these data were projected and adjusted from 2000 to 2015 at 5-year intervals by applying United Nations population growth estimates.

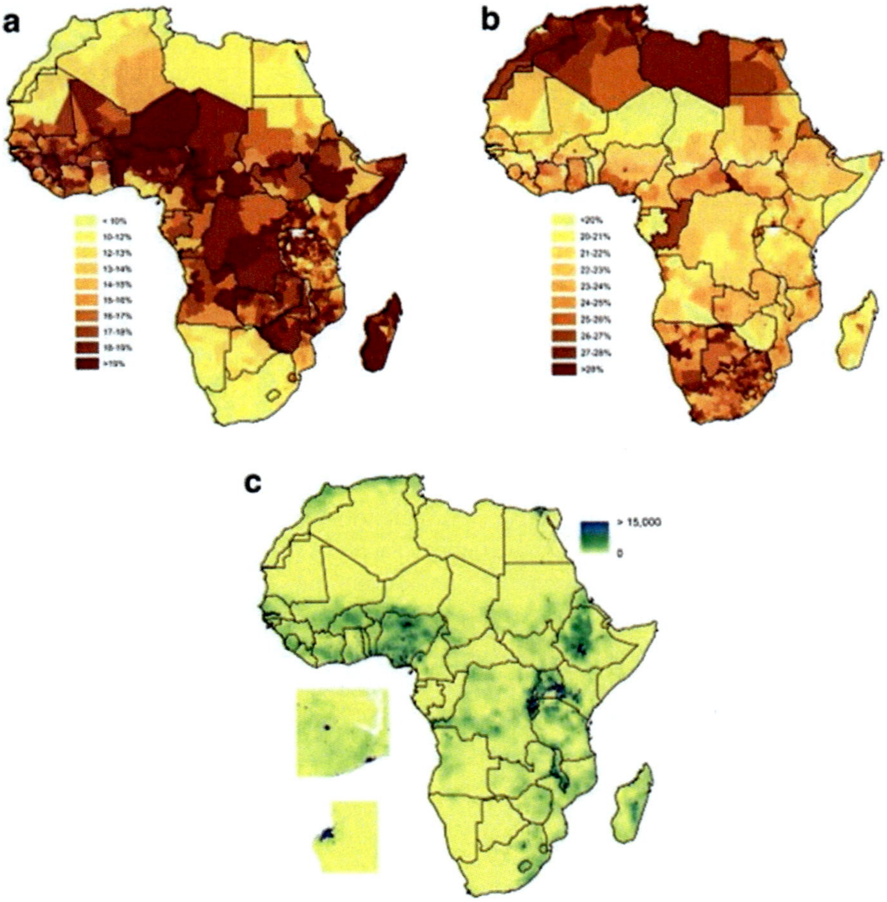

Fig. 2.5 Spatial demographic datasets for Africa. (**a**) Proportion of children under 5; (**b**) proportion of women of child-bearing age and (**c**) children under 5 in 2010 at 1 km resolution. (Adapted from Tatem et al. 2013)

As proof of concept, they further quantified how previously unaccounted for subnational demographic variation can impact important health and policy metrics, such as travel time to healthy facility and malaria transmission. By comparing these metrics generated using national level age structure data against the assembled subnational dataset described above, Tatem et al. found that modelled estimates could be as much as 100% discordant between each other, with as many as half of the study countries demonstrating greater than 10% discrepancies in estimates (Tatem et al. 2013). With crucial international development targets such as those associated with the Millennium Development Goals and Sustainable Development Goals tied to these estimates, in addition to national-level monitoring and surveillance funds, ensuring the accuracy of modelled estimates by capturing subnational demographic heterogeneity is vital to ensuring cost-effective decision making and policy efforts.

While this approach is useful in collating several data sources across a variety of countries, it is less useful for those countries where census data are not reliable and out of date. In the absence of such data sources, Alegana et al. developed a Bayesian hierarchical spatio-temporal model based framework to integrate georeferenced household survey data (such as those collected through the DHS) with a range of spatial data to predict age and gender structure at a 1 km by 1 km spatial resolution in Nigeria (Alegana et al. 2015). Similar to the bottom-up population mapping approach described above, this approach links geo-located cluster surveys with ancillary satellite data such as land cover, travel time to major settlements, night-time lights and vegetation index, to predict age structure in un-sampled locations, while statistically accounting for spatial autocorrelation. Overall, they found that these freely available high-resolution datasets (namely, land cover, night-time lights, travel time to major settlements and vegetation index) were strong predictors for estimating the proportion of the population under 5 years old. Because this framework utilizes only freely available household survey data, in combination with freely available remotely sensed data, this method fills an important niche for modelling high-resolution population distributions in the absence of reliable and modern census data. This work has important ramifications in health and policy metrics, including malaria surveillance and child and maternal mortality prevention, and has been used in vaccination planning and outreach in Nigeria through the Nigeria Vaccination Tracking System (vts.econg.org).

Estimating Live Births and Pregnancies

Building off previous techniques described above in mapping age and sex structure, Tatem et al. produced subnational datasets depicting women of childbearing age by 5-year intervals and number of live births and pregnancies (Tatem et al. 2014). To do this, Tatem and colleagues compiled age specific fertility rates (ASFR) from national surveys such as the DHS and Multiple Indicator Cluster Survey (MICS), disaggregated by subnational region and urban versus rural rates. These ASFR estimates were then combined with data on women of childbearing age, produced via the methods outlined above, to generate gridded estimates of live births (Fig. 2.6).

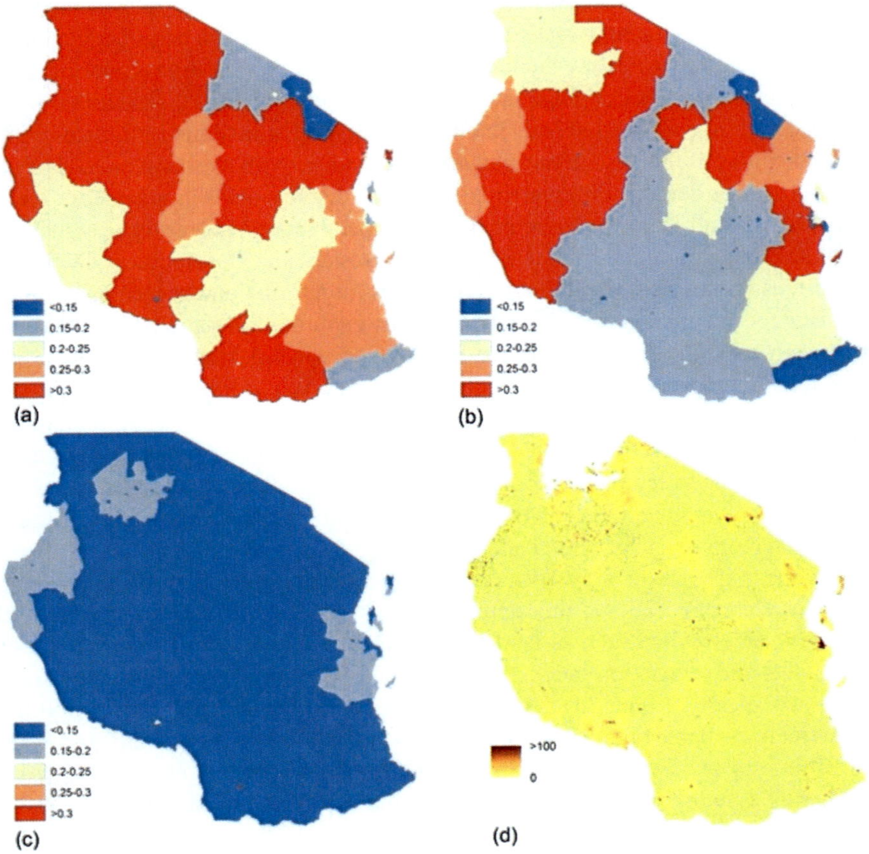

Fig. 2.6 Age-specific fertility rates among those aged (**a**) 20–24, (**b**) 30–34 and (**c**) 40–44, plus estimated live births per 100 × 100 m, Tanzania, 2012. (Adapted from Tatem et al. 2014)

Lastly, these estimates of live births were combined with data on abortions, stillbirths and miscarriages provided through the Guttmacher Institute to produce estimates of pregnancies per grid cell. These outputs are crucial in planning maternal and newborn health interventions and resource allocation, and were used in the United Nations Population Division's State of the World's Midwifery report for 2014 (UNFPA et al. 2014).

Quantifying Human Mobility

Since the nineteenth century, the volume of human mobility has grown, making crossing the globe easier than ever before (Tatem 2014). With this growth in human movement, pathogen spread is inevitable, and knowing how populations move, in addition to where they are distributed and who comprises them, is key for targeting interventions and allocating resources appropriately. Without accounting for this connectivity, global efforts towards eliminating disease spread such as malaria can be hampered, necessitating information on how people travel when designing eradication strategies (Sorichetta et al. 2016).

Towards this, anonymized call data records (CDRs) obtained via mobile phone SMS messages and calls are an increasingly useful tool in mapping human movement patterns at both a high spatial and temporal resolution, oftentimes consisting of millions of records spanning years. However, these data are particularly proprietary and hard to obtain due to extensive confidentiality concerns and network agreements. Using a mobile phone dataset for Kenya consisting of nearly 15 million users, Wesolowski et al. examined the extent to which census-derived migration data represented observed human movement as collected through CDRs (Wesolowski et al. 2013). National population and household survey census data represent the most widely used form of tracking human movement, and therefore potentially offer a robust and easily obtained form of data to quantify connectivity. Concerns exist, however, as to whether such data capture frequent, short-term movement, which is captured well by vast CDR datasets. Comparing these datasets, Wesolowski and colleagues found strong correlation between inbound, outbound and between-county movements (Fig. 2.7), with similar distributions of trip durations between the two datasets. These findings suggest that freely available census migration data may prove to be a useful proxy in approximating movement patterns across temporal scales, when larger and more fine-scale mobile phone datasets are not obtainable (Wesolowski et al. 2013).

Despite being difficult to obtain, CDRs contain both spatial and temporal information at unprecedented scales, representing a powerful tool in mapping fine-scale human mobility. These methods are therefore valuable in mapping population flows weekly, monthly, seasonally, or otherwise, with important ramifications in pathogen spread. For example, Ruktanonchai et al. used population flows to map malaria sources and sinks in the context of malaria elimination in Namibia (Ruktanonchai et al. 2016). Using a mobile phone dataset representing 9 billion communications from 1.2 million users in Namibia, they developed a mathematical model estimating malaria pathogen transmission sources and sinks, while incorporating observed human mobility patterns (Fig. 2.8). Specifically, they used prevalence estimates from 2009 with mobility patterns modelled using CDR data to identify pre-elimination malaria foci throughout Namibia. They further compared predicted transmission foci with post-2009 incident maps, and demonstrated that the spatial distribution of malaria cases shifted as malaria burden declined throughout the county. They suggest that spatial estimates modelling malaria prevalence and focal points must critically account for human movement and pathogen spread, particularly in the context of seasonal transmission, as evidenced by this framework. These findings have important policy implications in directing elimination and eradication efforts, particularly in near-elimination settings when malaria cases become more sparse.

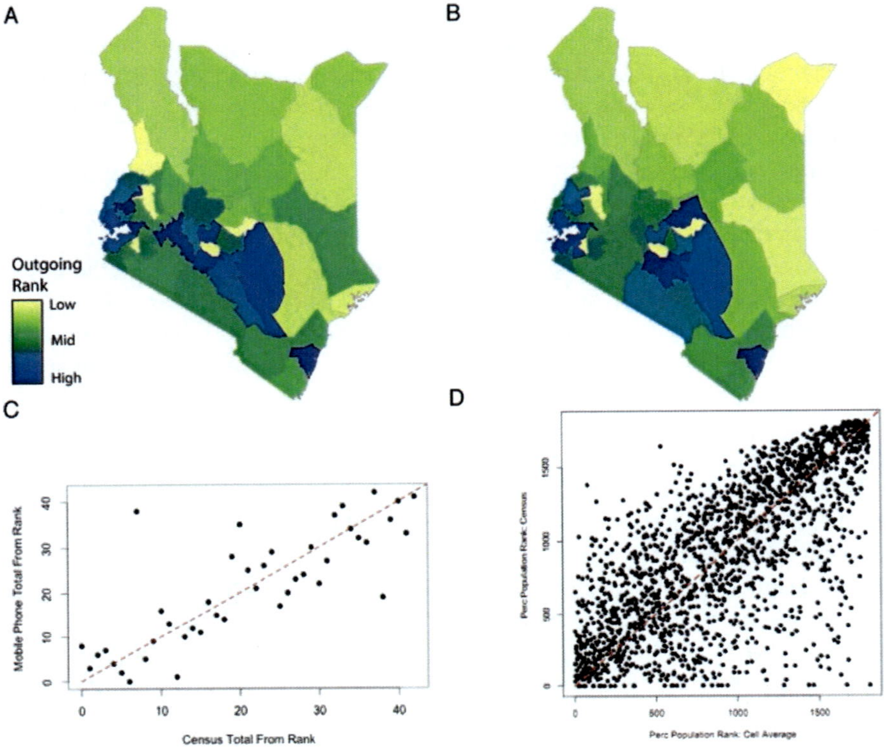

Fig. 2.7 (**a**) Human movement derived from mobile phone data; (**b**) migration data derived from census; (**c**) comparison of actual movement values between (**a**) and (**b**); and (**d**) percentage of population moving between all possible pairs of counties. (Adapted from Wesolowski et al. 2013)

Future Directions and Recommendations

Data visualization through mapping techniques represents a powerful tool for epidemiologists and demographers, particularly as the need for spatial investigation at a subnational scale becomes more prominent on the global health development agenda. Despite this, most research within public health does not incorporate spatial analysis (Auchincloss et al. 2012), and in fields such as maternal and newborn health, challenges persist in building analytic capacity among resource-poor settings (Ebener et al. 2015). Further, while several datasets exist such as census data, census microdata and national survey data, these are scattered across disparate sources and can be difficult to obtain, preventing comparison across time and space.

Towards this, Tatem et al. outline recommendations and future directions in improving spatial demographic datasets (Tatem et al. 2012). Firstly, they note that spatial and temporal population projections are lacking at a subnational level, even amongst international development agencies, such as the United Nations Population Division. While there are a few countries undertaking these projections at the

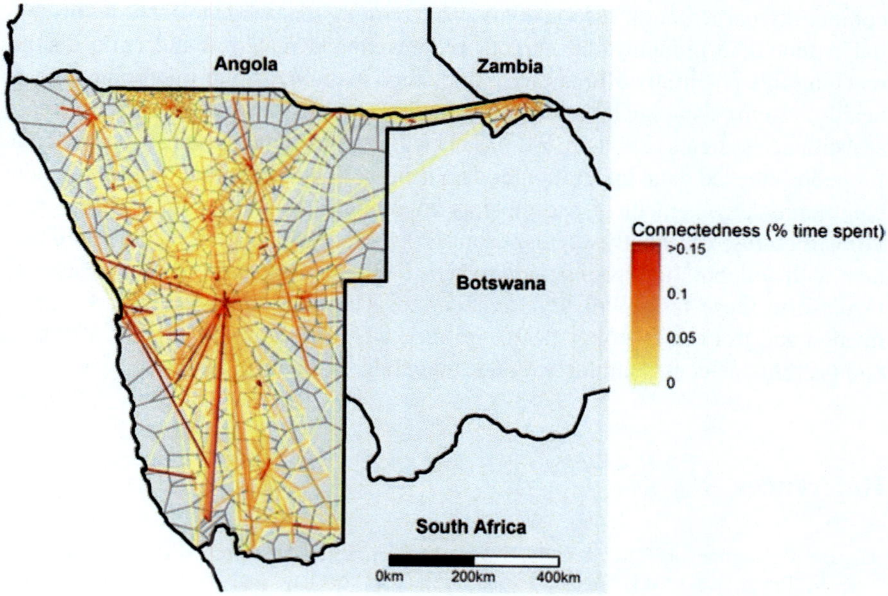

Fig. 2.8 Connectivity between patches derived from mobile phone data in Namibia, representing proportion of time spent between patches (i.e. how much time a resident in one patch spends in another patch). (Adapted from Ruktanonchai et al. 2016)

national level, such as India and China, these projections should be scaled up to include other developing countries to allow for more accurate population estimations and resource planning. The current methods used to project these estimates, however, are subject to sources of bias and model uncertainty, which Tatem et al. note must be better quantified and communicated. While efforts to map disease risk have handled the quantification and propagation of this uncertainty through Bayesian posterior distributions (Gething et al. 2015), less research has addressed understanding and visualizing how uncertainty propagates throughout spatial demographic datasets. This uncertainty might be as a result of outdated census data or the size of input administrative units as compared to output population grid cells, for example (Tatem et al. 2012). Without accounting for these drawbacks, spatial demographic datasets and corresponding population and health metrics may be substantially limited by unquantified sources of uncertainty propagating in unpredictable ways.

Finally, there is further a need for greater analytic capacity and data resources, particularly within resource poor settings (Ebener et al. 2015; Molla et al. 2017). Molla et al. outline future directions and policy recommendations for improving the use of GIS within the field of maternal and newborn health, but with broader applications towards health policy in developing countries. They categorize challenges and recommendations into three broad themes: (1) ancillary geospatial and maternal and newborn health data sources; (2) technical and human resources needs and (3) community participation (Molla et al. 2017). Noting that while GIS holds substantial potential in reducing the number of preventable maternal and child deaths, reaching this potential will require high quality spatial data and improving current health systems data, such as vital registration statistics and standardized disease surveillance systems (Rushton 2003). This will require a quality improvement cycle for georeferenced data, integration of data from both the public and private domain, community involvement in participatory mapping efforts to enhance spatial data infrastructures, and finally, regular monitoring and evaluation of inequities in tandem with collaboration from policy makers. Building in-country technical capacity to perform these tasks will be crucial in ensuring recommendations are implemented and in-country policy needs are met, which in turn will support efforts to end preventable deaths among women and children (Molla et al. 2017).

References

Alegana, V. A., Atkinson, P. M., Pezzulo, C., Sorichetta, A., Weiss, D., Bird, T., Erbach-Schoenberg, E., & Tatem, A. J. (2015). Fine resolution mapping of population age-structures for health and development applications. *Journal of the Royal Society Interface, 12*, 20150073. https://doi.org/10.1098/rsif.2015.0073.

Auchincloss, A. H., Gebreab, S. Y., Mair, C., & Roux, A. V. D. (2012). A review of spatial methods in epidemiology, 2000–2010. *Annual Review of Public Health, 33*, 107–122. https://doi.org/10.1146/annurev-publhealth-031811-124655.

Banerjee, S. (2016). Spatial data analysis. *Annual Review of Public Health, 37*, 47–60. https://doi.org/10.1146/annurev-publhealth-032315-021711.

Bhutta, Z. A., & Reddy, K. S. (2012). Achieving equity in global health: So near and yet so far. *JAMA, 307*, 2035–2036. https://doi.org/10.1001/jama.2012.4659.

de Montjoye, Y.-A., Hidalgo, C. A., Verleysen, M., & Blondel, V. D. (2013). Unique in the crowd: The privacy bounds of human mobility. *Scientific Reports, 3*, 1376. https://doi.org/10.1038/srep01376.

Deichmann, U., Balk, D., & Yetman, G. (2001). *Transforming population data for interdisciplinary usages: From census to grid*. Palisades, NY: NASA Socioeconomic Data and Applications Center (SEDAC), Columbia University.

Ebener, S., Guerra-Arias, M., Campbell, J., Tatem, A. J., Moran, A. C., Amoako Johnson, F., Fogstad, H., Stenberg, K., Neal, S., Bailey, P., Porter, R., & Matthews, Z. (2015). The geography of maternal and newborn health: The state of the art. *International Journal of Health Geographics, 14*, 19. https://doi.org/10.1186/s12942-015-0012-x.

Gething, P., Tatem, A., Bird, T., & Burgert, C. (2015). *Creating spatial interpolation surfaces with DHS data* (No. SAR 11). Rockville, MD: ICF International.

Hay, S. I., Noor, A. M., Nelson, A., & Tatem, A. J. (2005). The accuracy of human population maps for public health application. *Tropical Medicine & International Health, 10*, 1073–1086. https://doi.org/10.1111/j.1365-3156.2005.01487.x.

Mennis, J. (2003). Generating surface models of population using dasymetric mapping. *The Professional Geographer, 55*, 31–42. https://doi.org/10.1111/0033-0124.10042.

Molla, Y. B., Rawlins, B., Makanga, P. T., Cunningham, M., Ávila, J. E. H., Ruktanonchai, C. W., Singh, K., Alford, S., Thompson, M., Dwivedi, V., Moran, A. C., & Matthews, Z. (2017). Geographic information system for improving maternal and newborn health: Recommendations for policy and programs. *BMC Pregnancy and Childbirth, 17*, 26. https://doi.org/10.1186/s12884-016-1199-y.

Ruktanonchai, N. W., DeLeenheer, P., Tatem, A. J., Alegana, V. A., Caughlin, T. T., Erbach-Schoenberg, E. Z., Lourenço, C., Ruktanonchai, C. W., & Smith, D. L. (2016). Identifying malaria transmission foci for elimination using human mobility data. *PLOS Computational Biology, 12*, e1004846. https://doi.org/10.1371/journal.pcbi.1004846.

Rushton, G. (2003). Public health, GIS, and spatial analytic tools. *Annual Review of Public Health, 24*, 43–56. https://doi.org/10.1146/annurev.publhealth.24.012902.140843.

Sorichetta, A., Bird, T. J., Ruktanonchai, N. W., Erbach-Schoenberg, E. Z., Pezzulo, C., Tejedor, N., Waldock, I. C., Sadler, J. D., Garcia, A. J., Sedda, L., & Tatem, A. J. (2016). Mapping internal connectivity through human migration in malaria endemic countries. *Scientific Data, 3*, 160066. https://doi.org/10.1038/sdata.2016.66.

Stevens, F. R., Gaughan, A. E., Linard, C., & Tatem, A. J. (2015). Disaggregating census data for population mapping using random forests with remotely-sensed and ancillary data. *PLoS One, 10*, e0107042. https://doi.org/10.1371/journal.pone.0107042.

Tanser, F. C., & le Sueur, D. (2002). The application of geographical information systems to important public health problems in Africa. *International Journal of Health Geographics, 1*, 4. https://doi.org/10.1186/1476-072X-1-4.

Tatem, A. J. (2014). Mapping population and pathogen movements. *International Health, 6*, 5–11. https://doi.org/10.1093/inthealth/ihu006.

Tatem, A., & Linard, C. (2011). Population mapping of poor countries [WWW document]. *Nature, 474*, 36. https://doi.org/10.1038/474036d.

Tatem, A. J., Noor, A. M., von Hagen, C., Gregorio, A. D., & Hay, S. I. (2007). High resolution population maps for low income nations: Combining land cover and census in East Africa. *PLoS One, 2*, e1298. https://doi.org/10.1371/journal.pone.0001298.

Tatem, A. J., Adamo, S., Bharti, N., Burgert, C. R., Castro, M., Dorelien, A., Fink, G., Linard, C., John, M., Montana, L., Montgomery, M. R., Nelson, A., Noor, A. M., Pindolia, D., Yetman, G., & Balk, D. (2012). Mapping populations at risk: Improving spatial demographic data for infectious disease modeling and metric derivation. *Population Health Metrics, 10*, 8. https://doi.org/10.1186/1478-7954-10-8.

Tatem, A. J., Garcia, A. J., Snow, R. W., Noor, A. M., Gaughan, A. E., Gilbert, M., & Linard, C. (2013). Millennium development health metrics: Where do Africa's children and women of childbearing age live? *Population Health Metrics, 11*, 11. https://doi.org/10.1186/1478-7954-11-11.

Tatem, A. J., Campbell, J., Guerra-Arias, M., de Bernis, L., Moran, A., & Matthews, Z. (2014). Mapping for maternal and newborn health: The distributions of women of childbearing age, pregnancies and births. *International Journal of Health Geographics, 13*, 1–11. https://doi.org/10.1186/1476-072X-13-2.

UNFPA, ICM, WHO. (2014). *The state of the world's midwifery 2014: A universal pathway, a woman's right to health*. New York: United Nations Population Fund.

Varshney, K. R., Wei, D., Ramamurthy, K. N., & Mojsilović, A. (2015). Data challenges in disease response: The 2014 Ebola outbreak and beyond. *Journal of Data and Information Quality, 6*, 5. https://doi.org/10.1145/2742550.

Wardrop, N. A., Jochem, W. C., Bird, T. J., Chamberlain, H. R., Clarke, D., Kerr, D., Bengtsson, L., Juran, S., Seaman, V., & Tatem, A. J. (2018). Spatially disaggregated population estimates in the absence of national population and housing census data. *Proceedings of the National Academy of Sciences, 115*, 3529. https://doi.org/10.1073/pnas.1715305115.

Wesolowski, A., Buckee, C. O., Pindolia, D. K., Eagle, N., Smith, D. L., Garcia, A. J., & Tatem, A. J. (2013). The use of census migration data to approximate human movement patterns across temporal scales. *PLoS One, 8*, e52971. https://doi.org/10.1371/journal.pone.0052971.

Chapter 3
Geography of Disease Burden: Case Studies in Namibia and Eritrea

Victor A. Alegana and Peter M. Atkinson

Abstract Africa continues to experience the highest infectious disease burden despite an increase in investments. These include investments in malaria, HIV/AIDS, tuberculosis, as well as in communicable diseases. The global targets are to reduce the burden of these diseases through improved surveillance, prevention of outbreaks, effective case management, elimination and eventually, eradication. Achieving these targets, however, is limited by the poor geographic descriptions of the disease burden. Of the big five infectious disease burdens, malaria is the most advanced in terms of mapping its distribution. Malaria cartography has since formed the evidence-base for the design of many national malaria control programmes. This chapter focuses on malaria as an example, demonstrating its geographical descriptions. The availability of georeferenced malaria case data whether based on prevalence or incidence indicators has been used extensively in the mapping of geographical extents at national and sub-national scales. However, routine surveillance data is emerging as a valuable methodology of tracking burden in sub-Saharan Africa. A particular focus of this chapter is the use of routine national health systems surveillance data to describe, at a fine-scale, the distribution of malaria. However, routine data can be applied to the cartographic description of other diseases beyond malaria. The methodological aspects of burden estimation from routine surveillance platforms and cartography are highlighted.

Keywords Burden · Disease mapping · Routine data · Surveillance

V. A. Alegana (✉)
Population Health Unit, Kenya Medical Research Institute-Wellcome Trust Research Programme, Nairobi, Kenya

Nuffield Department of Medicine, Centre for Tropical Medicine and Global Health, John Radcliffe Hospital, University of Oxford, Oxford, UK

Geography and Environmental Science, University of Southampton, Southampton, UK
e-mail: valegana@kemri-wellcome.org

P. M. Atkinson
Geography and Environmental Science, University of Southampton, Southampton, UK

Faculty of Science and Technology, Lancaster University, Lancaster, UK

Institute of Geographic Sciences and Natural Resources Research,
Chinese Academy of Sciences, Beijing, China

P. T. Makanga (ed.), *Practicing Health Geography*, Global Perspectives on Health Geography, https://doi.org/10.1007/978-3-030-63471-1_3

Introduction

Thegeography of disease mapping in Africa stems back to 1951 following publication of atlas of diseases in Africa after the Second World War (Simmons et al. 1951). Historical attempts to eradicate malaria in Africa started mid-1950s during the Global Malaria Eradication Programme (GMEP) era. In the 1950s to 1960s, many African colonial governments developed crude national-level maps of malaria based on ecological zones and seasonality as part of the GMEP planning, for example, in Kenya (Butler 1959), Madagascar (Joncour 1956), Senegal (Lariviere et al. 1961), and Uganda (Mccrae 1975). The Failure of GMEP in Africa led to the resurgence of malaria through the 1970s and 1980s. Efforts to describe the geographical extent of malaria in Africa were resurrected in the 1990s. In 1996, the Mapping Malaria Risk in Africa/Atlas du Risqué de la Malaria en Afrique (MARA/ARMA), a collaboration between Africa research institutes, started to assemble data on malaria prevalence in sub-Saharan Africa (SSA) (Snow et al. 1996; Le Sueur et al. 1997). This was an initiative that started an assembly of data on malaria prevalence in Africa to used in malaria cartographic descriptions. Advances in computation and Geographic Information Systems (GIS) between the mid-1990s and 2000s has aided the development of robust malaria cartography including statistical description at national and sub-regional levels independently (Craig et al. 2007; Gemperli et al. 2006; Kazembe 2007; Noor et al. 2008, 2009, 2013b, 2014), and through the inception of the Malaria Atlas Project in the mid-2000s (Hay et al. 2004, 2008; Snow et al. 2005; Hay and Snow 2006; Guerra et al. 2007; Snow 2014).

Prevalence or incidence are two common indices that are now used frequently in the mapping of malaria (Macdonald 1950, 1957; Ray and Beljaev 1984). These indices provide epidemiological evidence of the spatial distribution of disease in the population. Prevalence represents the number of cases or infections at a given time (cross-section measure), while incidence represents the number of new cases arising over a specified period in the population (dynamic measure) (Fig. 3.1). Prevalence is usually stated as a rate (i.e. per fixed number in the population) while incidence is commonly expressed as the number of cases per 1000 population per year (Swaroop et al. 1966; Pull 1972). There are many reasons for describing the geography of these two metrics. Maps are useful tools to visualise the extent of a public health problem and for planning interventions. Maps can also be used as measurement tools to assess the impact of public health investments providing evidence on the success or failure of health interventions (Hay et al. 2013).

Since the 1990s, with advances in computation and software, maps of malaria prevalence and incidence are increasingly available at global and national scales. These maps, however, are presented with varying degrees of precision due to the wide variety of approaches used in their production. Variation in the cartographic description of prevalence and incidence in sub-Saharan Africa is also driven by the quality and quantity of data available. For malaria, countries with good surveillance systems utilise routine data without the requirement for modelling, e.g. Comoros and Sao Tome and Principe (Alegana et al. 2020). However, poor quality of the

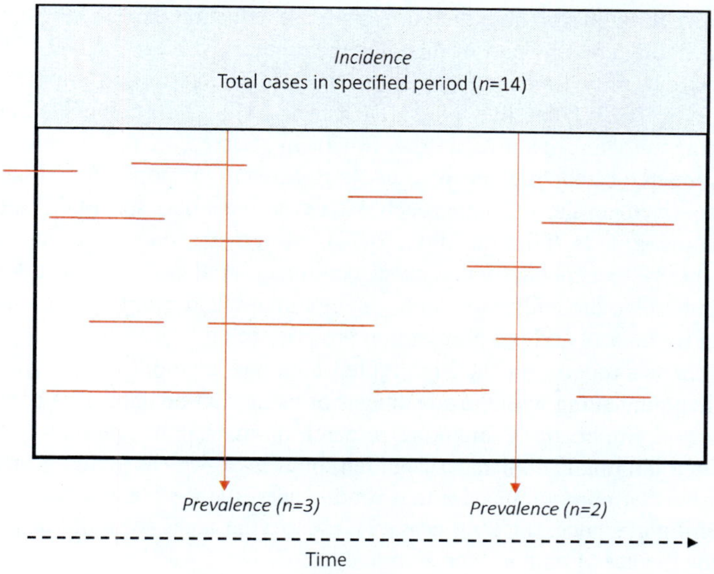

Fig. 3.1 Differences between prevalence and incidence of disease. Each horizontal red bar represents a case with the length of bar illustrating the duration of illness, e.g., fever

routine data require modelling to adjust for the use of health services, inconsistent data reporting, and climatic drivers of transmission. As a result, complex statistical modelling schemes have been developed for mapping disease (Diggle et al. 1998; Giorgi et al. 2015).

Model-based geostatistical (MBG) approaches combined with environmental variables (predictors) that support dynamic transmission and incidence are now used commonly to produce a gridded, fine spatial resolution estimates (Diggle et al. 1998). The advantage of MBG methods is the ability to harness the spatial and temporal dependencies in the observed data and environmental predictors. MBG also estimates the uncertainty associated with the predicted maps which are often defined in space and time. In practice, the generalised linear mixed class of models is used to connect the observed data to environmental predictors (Dalrymple et al. 2015; Alegana et al. 2016). The precision and accuracy of predictions can be evaluated via internal model parameters, or via exceedance probabilities (Giorgi et al. 2018), and by comparing to out of sample data. One source of uncertainty in passive surveillance systems such as the Health Management Information Systems (HMIS), is contributed by variation in health sector use by the population.

Several further issues impact our ability to describe the geography of disease burden. Firstly, as prevalence declines, increasingly large sampling at the community level is required. Disease biomarkers are included in surveys conducted every 3–4 years. The precision of various biomarkers in these community surveys therefore varies and may not always be optimal for monitoring and evaluation (Alegana et al. 2017). For malaria, as prevalence declines in sub-Saharan Africa, the use of surveillance through a combination of active case detection (ACD) and passive case detection (PCD) are now part of the Global Technical Strategy (WHO 2015). This method is currently used in Swaziland and a few countries in southern Africa (Hsiang et al. 2012; Dlamini et al. 2018). Reactive Case Detection (RACD) is also used during epidemics (Sturrock et al. 2013). In practice, PCD is labour intensive and is hampered by the high costs of tracking cases in the population. PCD should ideally complement the ACD approach. However, most data from PCD are unreliable and incomplete (Githinji et al. 2017). Moreover, some case data reported through this system are based on clinical examination rather than parasitology. With declining burden, the ability to identify symptomatic and asymptomatic infections is critical for control and pre-elimination programmes.

This chapter reports on the highlighted data and methodological advances in disease mapping along with the advantages of using routine data. This contribution has important implications for future research on malaria in line with a declining burden for traditionally high malaria burden countries as well as for low-transmission settings. Furthermore, emphasise that routine surveillance remains the foundation for gathering evidence, tracking progress, identifying areas for rapid response and promoting the use of data for decision making.

Disease Cartography from Routine Surveillance Systems

Role of Surveillance for Geographies of Disease

Surveillance started in the 1950s as part of GMEP and was used as a means of preventing re-emergence of disease (World Health Organization 2012a). According to the WHO, surveillance included the identification of infections, investigation, elimination of transmission and prevention as well as cure. Surveillance is a recommended intervention for tracking disease burden for targeting interventions. There are two broad areas concerned with determining the incidence of disease including the identification of cases (PCD) and elimination of the identified cases.

Introduction to Using PCD for Mapping

Innovative approaches now exist to harness PCD and, thus, complement ACD which is not yet adopted by much of sub-Saharan Africa. To properly utilise routine data, there is a need to establish the denominator, i.e. population covered by the health system (the catchment population). The methods now exist for capturing the febrile population using the healthcare system and combining this with fine spatial resolution population maps (Tatem 2014) to estimate catchment populations. Secondly, alongside improvements in HMIS data, for example, through District Health Information Systems (DHIS 2) (Karuri et al. 2014; Dehnavieh et al. 2018) statistical techniques can be used to model the spatial and temporal variability in incidence while at the same time accounting for the rate of health facility utilisation and incompleteness (Alegana et al. 2013). Such approaches incorporate ecological or environmental drivers to predict risk in receptive areas while at the same time quantifying the uncertainty associated with disease predictions (Noor et al. 2012, 2013a).

Overcoming Barriers in Mapping Using HMIS Data

Health facility data serve as indicators of the disease epidemiology amongst the populations they serve. As surveillance centres, health facilities are better barometers of changing disease landscapes than modelled snapshots of prevalence.

Despite methodological advances, HMIS data have been previously ignored for burden estimation because of incomplete reporting and variation in the population using public health sectors across sub-Saharan Africa (Battle et al. 2016; Alegana et al. 2018). This implies that cases recorded at the health facility often indicate only the '*tip of the iceberg*' of the actual burden. This variation in utilisation potentially introduces a bias in the estimation of disease burden. In addition, weak health systems in relation to the quality and quantity of data have in the past contributed to a general lack of confidence in the use of health facility data in sub-Saharan Africa.

HMIS, however, remain an important source of data for future disease mapping for several reasons. Firstly, the spatial distribution of health facilities is usually congruent to the population distribution (Fig. 3.2). Secondly, health facility case data are often collected in an ongoing manner (e.g. daily, weekly and monthly) (Mueller et al. 2011). The implication is that data are likely to have a temporal signal useful in identifying the seasonal dynamics of the disease. Thirdly, the coverage of a health facility catchment population often encompasses several villages, communities and sometimes the whole administrative region (e.g. district). This implies a wider geographic coverage of a single health facility in an HMIS than of a single village in a cross-sectional prevalence survey.

The Geography of the Denominator for Burden Estimation

Disease estimation based on health facility data requires a definition of the denominator (febrile population within the health facility catchment population). Thus, using health facility data for mapping incidence requires an adjustment for healthcare use, both in the public and private sectors. Utilisation has in the past been estimated from household surveys by quantifying the probability of public or private sector use (Stekelenburg et al. 2005; Noor et al. 2006). Such an approach is potentially beneficial in identifying the population not covered by healthcare systems. Previously, this has been characterised in GIS by defining a distance metric (or travel times) (Apparicio and Seguin 2006; Noor et al. 2006; Apparicio et al. 2008). At the second stage, the reported rates of use at the community level are

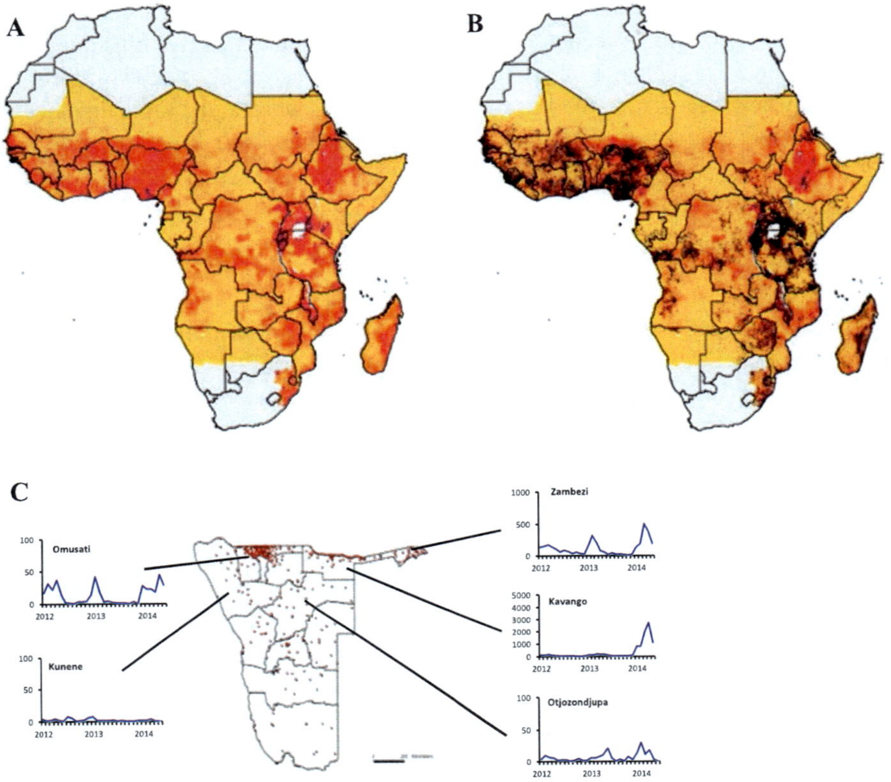

Fig. 3.2 (**a**) Population density map of sub-Saharan Africa, (**b**) the spatial distribution of health facilities superimposed on the population map (Maina et al. 2019), (**c**) an illustration of outpatient malaria cases from administrative areas in Namibia showing variation in the number and seasonal patterns (Alegana et al. 2013). These seasonal average trends have not been adjusted based on total facility reporting rate. In this case, the Namibia case reporting rate was greater than 90% at the regional level

modelled as a function of travel time or distance to define a utilisation probability index (Alegana et al. 2012) (Fig. 3.3). The probabilistic estimate is useful for burden estimation because patients located far from a health facility are less likely to be treated in a formal care setting. It is then possible to estimate a population coverage indicator as well as hard catchment boundaries (e.g. for probability >40%). An example of this approach has been used in Namibia to zone catchment areas and estimate the age-structured catchment population (Alegana et al. 2016).

Environmental Drivers of Geographical Risk

Disease burden mapping generally requires a statistical model with a suitable combination of environmental variables (covariates) to predict incidence or prevalence. Several covariates have been shown to drive disease dynamics and transmission. It

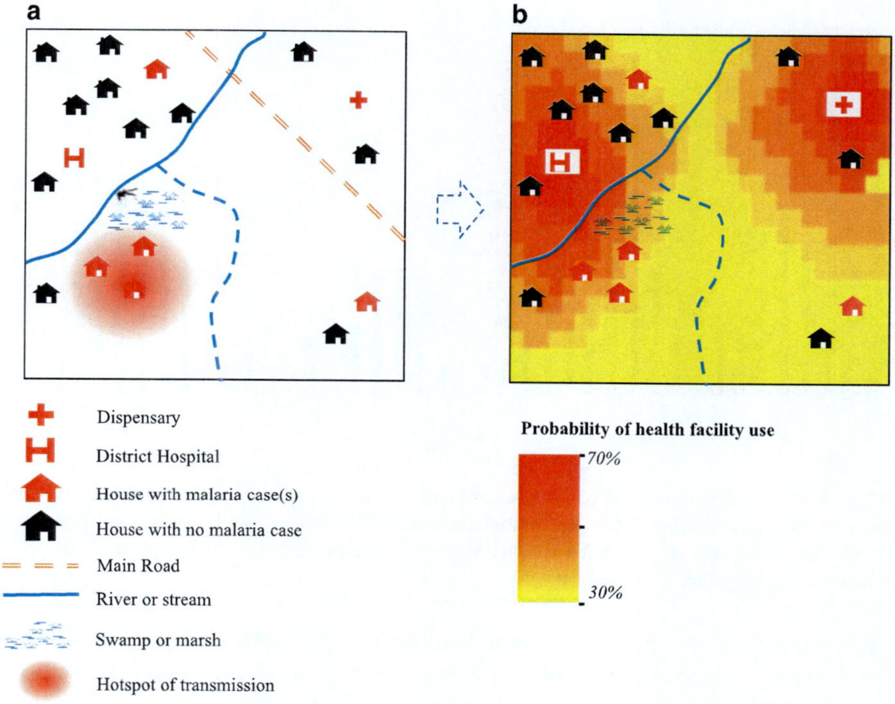

Fig. 3.3 (**a**) Shows an example of malaria landscape showing transmission aspects (mosquito habitats). Often environmental suitability drives transmission and the location of a hotspot could be far from the nearest health facility. (**b**) A representation of the probability of seeking treatment at health facilities. Often probability of use within the health facility catchment area reduces with geographic distance as well as other socio-demographic factors. (Adopted from Alegana et al. 2016)

is important to select a biologically plausible set of covariates related to disease based on some criterion to achieve parsimony. This is because using many covariates may result in over-fitting or introduce multicollinearity (Babyak 2004). Thus, preliminary selection of a set of covariates that best describes the response is a widely accepted exercise in statistical modelling of burden (Murtaugh 2009).

For malaria mapping, environmental variables affect the development and survival of the malaria parasite as well as the malaria vector (Molineaux et al. 1988). Examples of these include the monthly rate of precipitation, temperature, vegetation cover, aridity and urbanisation (Craig et al. 1999; Guerra et al. 2008). Figure 3.4 shows an example of satellite remotely sensed covariates plotted against PCD in Eritrea. These include precipitation, minimum temperature, maximum temperature and mean temperature, the normalised difference vegetation index (NDVI) and the enhanced vegetation index (EVI). The vegetation indices are derived from MODerate-resolution Imaging Spectroradiometer (MODIS) sensor imagery; produced after removing heavy aerosols through atmospheric correction, elimination of shadows and clouds and correcting to bidirectional reflectance (Huete et al. 2002). The mean monthly gridded temperature estimates were downloaded from the WorldClim repository at

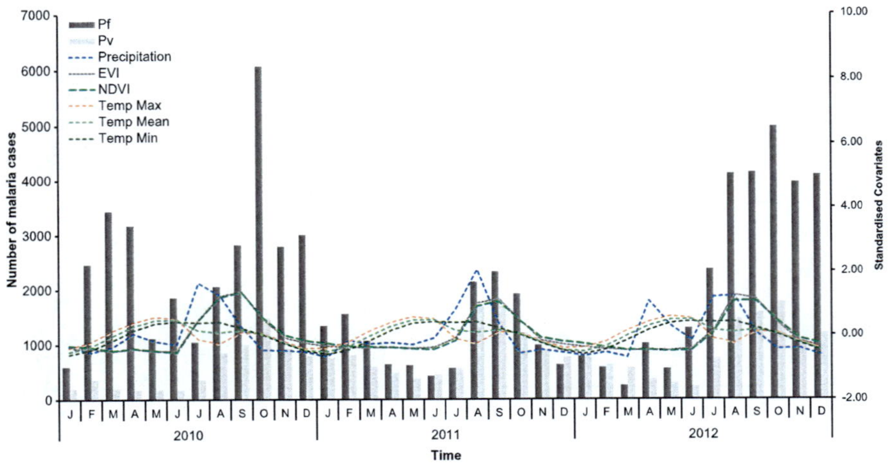

Fig. 3.4 Seasonal monthly plot of the observed malaria cases (green bars) from 2010 to 2012 with the dark grey representing *P. falciparum* malaria cases and light grey the *P. vivax* cases. The magnitude for the cases is shown on the primary vertical axis. Covariates (secondary vertical axis) are plotted as dashed lines

approximately 1 km spatial resolution (0.000833° × 0.000833°). These gridded estimates were produced from long-term climate observations for the period 1950–2000, interpolated using smoothing spline algorithms. Precipitation data were obtained from the Tropical Rainfall Measuring Mission sensor (TRMM 3B43 product) that combines ground observations and satellite sensor data to generate gridded rainfall estimates at approximately 0.25° × 0.25° spatial resolution (Huffman et al. 2007). TRMM 3B43 is a gridded mean monthly average product of precipitation rate in mm h^{-1}.

Example of Mapping PCD from HMIS

Spatial regression models are common in disease mapping (Bernardinelli et al. 1997; Clements et al. 2006; Schrödle and Held 2010, 2011). Two common approaches involve smoothing of disease rates in space applying small-area estimation methods (conditional autoregressive (CAR)) and the interpolation via geostatistical approaches (Banerjee et al. 2004). The CAR framework involves spatial smoothing between administrative areas (e.g. districts) (Besag et al. 1991). The level of smoothing is controlled via modelling parameters. A suitable smoothing approach should take into consideration the arrangement of spatial units to yield optimal spatial variation. A general problem common to this approach, however, relates to a change in the statistical outputs as a result of a change in the shape or size of the geographic unit, the *modifiable areal unit problem* (MAUP). Hierarchical modelling aims to mitigate some aspects of MAUP.

An example of the use of a hierarchical Bayesian model applied to smooth monthly malaria incidence at the district level for case data is shown based on data in Fig. 3.4. The numerator is presented as the sum of cases recorded at the facility (include both confirmed case through parasitology diagnosis and clinical diagnosis). The denominator was derived from the population-weighted catchments representing all-age febrile case risk at each health facility. An adjustment is made to clinical case diagnosis using the slide positivity rate at the facility for the numerator. For the denominator, adjustment is necessary for reporting rate and health facility use (Fig. 3.5). This modelling example was conducted using facility-level data. Thus, a facility-level random effect was incorporated to allow for variation between facilities at the district level as well as a seasonal trend. Such an approach improves smoothing and estimation. To deal with incomplete reporting missing data months were imputed as 'NAs'. Random effects were incorporated at the district and regional levels. Non-linear parametric smoothing functions were used for the covariates rather than an assumption of linearity (constant) (Fahrmeir and Knorr-Held 2000) (Fig. 3.6).

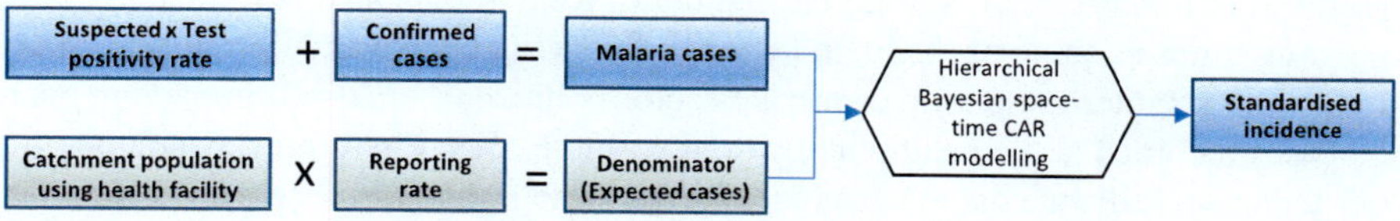

Fig. 3.5 Schematic diagram showing the general modelling framework. The Test Positivity Rate (TPR) is defined based on the testing proportion at the health facility while the estimation of the catchment population is based on geographic access modelling

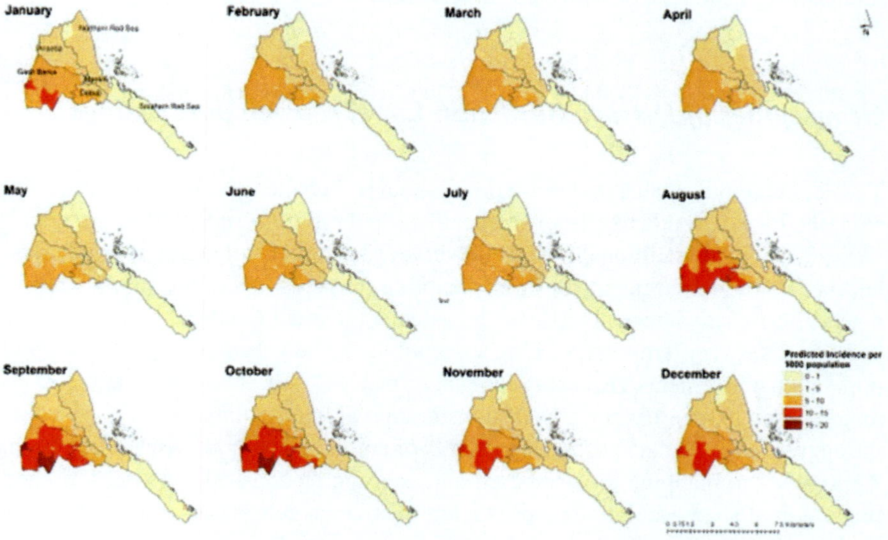

Fig. 3.6 An example of monthly maps of the incidence of *P. falciparum* per 1000 population in Eritrea using a Bayesian spatio-temporal Poisson model. Districts with low risk are classified as <5 cases per 1000 population) and moderate risk with >5 cases per 1000 population. Data were from a 3-year time-series (2010–2012) of malaria cases from the HMIS

Discussion

Routine Surveillance for Mapping Disease Burden

The chapter aimed to highlight issues around the use of these data in SSA, challenges and examples of methods deployed to map routine surveillance data. HMIS coordinates the routine acquisition of data from health facilities (public and private) and compilation of these data (e.g. cases) through the district, regional and national levels (Abouzahr et al. 2007; Boerma and Stansfield 2007). Such data form an integral part of healthcare delivery and are useful for planning, resource allocation and disease monitoring. In reality, however, HMIS are often incomplete in many African countries as outlined in this chapter and the utilisation of health facilities is not uniform. Some of the factors contributing to low facility utilisation include the availability of health services, financial factors, geographic access and waiting times at facilities (Breman 2001). Studies carried out in Kenya suggested cost, distance and opening times as some of the main factors influencing choice and decisions to seek treatment in either the public or private sector (Chuma et al. 2010) impacting data on cases recorded at the health facility and within HMIS. Therefore, specific methods using surveillance data to produce disease cartography are necessary to smooth estimates of incidence and adjust for sporadic reporting and utilisation by the population. These were demonstrated in this chapter alongside accounting for environmental variables when estimating incidence. Mapping disease incidence is important to the various national health programmes for resource allocation and provides useful insights in carrying out targeted surveillance.

Geographies of Disease Burden in Low-Transmission Settings

The declining prevalence of disease presents several challenges. With low transmission, the disease tends to cluster in specific population '*hotspots*' (Bousema et al. 2012, 2016). The traditional household surveys become challenging to implement because of the requirement for large sampling and cross-sectional surveys fail to detect short-term changes in disease prevalence (at small temporal scales). This is because cases vary temporally, being susceptible to changes in climate, ecology and population movements (Erbach-Schoenberg et al. 2016). The cartographic challenge is then to identify hotspots of transmission at fine spatial resolution based on the aggregated case data observed passively or combined with active case detection. Approaches to mapping disease based on cases aggregated at the district level and prediction of spatio-temporal maps at a fine spatial resolution can be used (Alegana et al. 2016) (Fig. 3.7). The approach improves the ability to characterise hotspots at the fine spatial resolution and can be used to target resources to specific local populations. This targeting can be cost-effective where the population distribution is sparse and further surveillance can be limited to specific local areas.

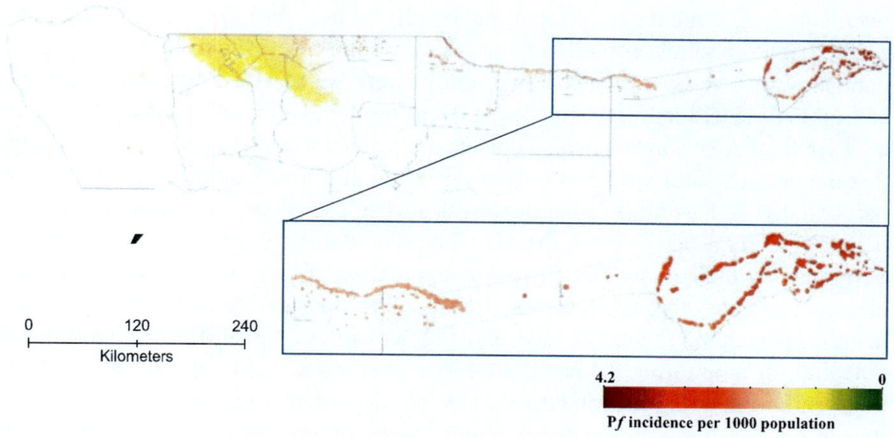

Fig. 3.7 Example of a fine-resolution map of incidence for Namibia based on the data in Fig. 3.2c. Map produced only for the endemic northern regions of Namibia

Challenges and Opportunities for Cartography for Elimination

Progress in identifying symptomatic cases within the population has important implications for asymptomatic case detection. Both PACD and RACD will benefit from the improved mapping of passively detected cases at fine spatial resolution. Improvement in routine data quality is likely to enhance malaria cartography. A different challenge exists in areas where multiple malaria parasites co-infect (Cotter et al. 2013). Most of the approaches outlined for disease cartography often focus on one parasite species. For example, there has been some progress in mapping other malaria parasites on the continent such as *P. vivax* (Battle et al. 2019). There is increasing evidence of *Pv* distribution (Twohig et al. 2019). More effective approaches need to be developed for mapping co-infections (Commons et al. 2019). The challenges posed by *P. vivax* are considerable due to the biological characteristics (Mueller et al. 2009). *P. vivax* exhibits a dormant liver stage responsible for most relapses up to weeks or months after an initial attack (White 2011). This complicates the ability to detect and apply suitable cartographic approaches to the asymptomatic *Pf* and *Pv* co-infections within the population.

Conclusion

The last decade has seen a transformation in Health Management Information System (HMIS) data in Africa. Two key advances include data digitisation of data through DHIS2 and Firstly, the ability to define malaria-specific morbidity presenting to the health facilities through Test. Treat. Track (T3) initiative (World Health Organization 2012b). The potential benefit of this transformation cannot be over-stated since the data represent the entirety of the presenting cases in national public health systems in participating African countries. Moreover, most African nations now have operational digital and georeferenced HMIS, meaning that the ensemble of HMIS represents a powerful lens through which to assess the health of the people of Africa as a whole. The data are subject to some biases, most notably that the public health system is only a part of the full health system, albeit a major part and that under-utilisation of the health system can occur at alarming rates, particularly in rural areas. Nevertheless, Bayesian statistical approaches have been developed by the authors that allow for suppression of these biases when mapping disease incidence through space and time. With appropriate Bayesian statistical handling, including the use of environmental covariates, the HMIS data have great potential for monitoring the health of Africa over space and time and for targeting interventions in both space and time. They have a particular utility for low endemicity settings, or in pre-elimination settings, where prevalence of disease is low and clustered in hotspots. In such settings, active case detection is extremely inefficient to the point of being unusable, and passive case detection, as afforded by the HMIS, can be invaluable for residual or emerging hotspot detection. We hope that this chapter will lead to greater awareness of the potential of African HMIS for and the space-time statistical techniques that allow their proper and principled use.

Acknowledgements *Funding statement*: Victor Alegana is funded through a Wellcome Trust Training Fellowship (#211208).

References

Abouzahr, C., Adjei, S., & Kanchanachitra, C. (2007). From data to policy: Good practices and cautionary tales. *The Lancet, 369*, 1039–1046.

Alegana, V., Wright, J., Petrina, U., Noor, A., Snow, R., & Atkinson, P. (2012). Spatial modelling of healthcare utilisation for treatment of fever in Namibia. *International Journal of Health Geographics, 11*, 6.

Alegana, V. A., Atkinson, P. M., Wright, J. A., Kamwi, R., Uusiku, P., Katokele, S., Snow, R. W., & Noor, A. M. (2013). Estimation of malaria incidence in northern Namibia in 2009 using Bayesian conditional-autoregressive spatial-temporal models. *Spatial and Spatio-temporal Epidemiology, 7*, 25–36.

Alegana, V. A., Atkinson, P. M., Lourenço, C., Ruktanonchai, N. W., Bosco, C., Erbach-Schoenberg, E. Z., Didier, B., Pindolia, D., Le Menach, A., Katokele, S., et al. (2016). Advances in mapping malaria for elimination: Fine resolution modelling of Plasmodium falciparum incidence. *Scientific Reports, 6*, 29628.

Alegana, V. A., Wright, J., Bosco, C., Okiro, E. A., Atkinson, P. M., Snow, R. W., Tatem, A. J. & Noor, A. M. (2017). Malaria prevalence metrics in low- and middle-income countries: an assessment of precision in nationally-representative surveys. *Malaria Journal, 16*, 475.

Alegana, V. A., Maina, J., Ouma, P. O., Macharia, P. M., Wright, J., Atkinson, P. M., Okiro, E. A., Snow, R. W., & Tatem, A. J. (2018). National and sub-national variation in patterns of febrile case management in sub-Saharan Africa. *Nature Communications, 9*, 4994.

Alegana, V. A., Okiro, E. A., & Snow, R. W. (2020). Routine data for malaria morbidity estimation in Africa: Challenges and prospects. *BMC Medicine, 18*, 121.

Apparicio, P., & Seguin, A.-M. (2006). Measuring the accessibility of services and facilities for residents of public housing in Montreal. *Urban Studies, 43*, 187–211.

Apparicio, P., Abdelmajid, M., Riva, M., & Shearmur, R. (2008). Comparing alternative approaches to measuring the geographical accessibility of urban health services: Distance types and aggregation-error issues. *International Journal of Health Geographics, 7*, 7.

Babyak, M. A. (2004). What you see may not be what you get: A brief, nontechnical introduction to overfitting in regression-type models. *Psychosomatic Medicine, 66*, 411–421.

Banerjee, S., Carling, P. B., & Gelfand, A. E. (2004). *Hierarchical modeling and analysis for spatial data*. London: Chapman & Hall/CRC.

Battle, K. E., Bisanzio, D., Gibson, H. S., Bhatt, S., Cameron, E., Weiss, D. J., Mappin, B., Dalrymple, U., Howes, R. E., Hay, S. I., et al. (2016). Treatment-seeking rates in malaria endemic countries. *Malaria Journal, 15*, 1–11.

Battle, K. E., Lucas, T. C. D., Nguyen, M., Howes, R. E., Nandi, A. K., Twohig, K. A., Pfeffer, D. A., Cameron, E., Rao, P. C., Casey, D., et al. (2019). Mapping the global endemicity and clinical burden of *Plasmodium vivax*, 2000-17: A spatial and temporal modelling study. *The Lancet, 394*, 332–343.

Bernardinelli, L., Pascutto, C., Best, N. G., & Gilks, W. R. (1997). Disease mapping with errors in covariates. *Statistics in Medicine, 16*, 741–752.

Besag, J., York, J., & Mollie, A. (1991). Bayesian image restoration, with two applications in spatial statistics. *Annals of the Institute of Statistical Mathematics, 43*, 1–59.

Boerma, J. T., & Stansfield, S. K. (2007). Health statistics now: Are we making the right investments? *The Lancet, 369*, 779–786.

Bousema, T., Griffin, J. T., Sauerwein, R. W., Smith, D. L., Churcher, T. S., Takken, W., Ghani, A., Drakeley, C., & Gosling, R. (2012). Hitting hotspots: Spatial targeting of malaria for control and elimination. *PLoS Medicine, 9*, e1001165.

Bousema, T., Stresman, G., Baidjoe, A. Y., Bradley, J., Knight, P., Stone, W., Osoti, V., Makori, E., Owaga, C., Odongo, W., et al. (2016). The impact of hotspot-targeted interventions on malaria transmission in Rachuonyo South District in the Western Kenyan highlands: A cluster-randomized controlled trial. *PLoS Medicine, 13*, e1001993.

Breman, J. G. (2001). The ears of the hippopotamus: Manifestations, determinants, and estimates of the malaria burden. *The American Journal of Tropical Medicine and Hygiene, 64*, 1–11.

Butler, R. (1959). *Atlas of Kenya: A comprehensive series of new and authenticated maps prepared from the national survey and other governmental sources with gazetteer and notes on pronunciations and spelling*. Nairobi: Survey of Kenya.

Chuma, J., Okungu, V., & Molyneux, C. (2010). Barriers to prompt and effective malaria treatment among the poorest population in Kenya. *Malaria Journal, 9*, 144.

Clements, A. C., Lwambo, N. J., Blair, L., Nyandindi, U., Kaatano, G., Kinung'hi, S., Webster, J. P., Fenwick, A., & Brooker, S. (2006). Bayesian spatial analysis and disease mapping: Tools to enhance planning and implementation of a schistosomiasis control programme in Tanzania. *Tropical Medicine & International Health, 11*, 490–503.

Commons, R. J., Simpson, J. A., Thriemer, K., Hossain, M. S., Douglas, N. M., Humphreys, G. S., Sibley, C. H., Guerin, P. J., & Price, R. N. (2019). Risk of *Plasmodium vivax* parasitaemia after *Plasmodium falciparum* infection: A systematic review and meta-analysis. *The Lancet Infectious Diseases, 19*, 91–101.

Cotter, C., Sturrock, H. J. W., Hsiang, M. S., Liu, J., Phillips, A. A., Hwang, J., Gueye, C. S., Fullman, N., Gosling, R. D., & Feachem, R. G. A. (2013). The changing epidemiology of malaria elimination: New strategies for new challenges. *The Lancet, 382*, 900–911.

Craig, M. H., Snow, R. W., & Le Sueur, D. (1999). A climate-based distribution model of malaria transmission in sub-Saharan Africa. *Parasitology Today, 15*, 105–111.

Craig, M. H., Sharp, B. L., Mabaso, M. L. & Kleinschmidt, I. (2007). Developing a spatial-statistical model and map of historical malaria prevalence in Botswana using a staged variable selection procedure. Int J Health Geogr, *6*, 44.

Dalrymple, U., Mappin, B., & Gething, P. (2015). Malaria mapping: Understanding the global endemicity of falciparum and vivax malaria. *BMC Medicine, 13*, 140.

Dehnavieh, R., Haghdoost, A., Khosravi, A., Hoseinabadi, F., Rahimi, H., Poursheikhali, A., Khajehpour, N., Khajeh, Z., Mirshekari, N., Hasani, M., et al. (2018). The District Health Information System (DHIS2): A literature review and meta-synthesis of its strengths and operational challenges based on the experiences of 11 countries. *Health Information Management Journal, 48*, 62–75.

Diggle, P., Tawn, J., & Moyeed, R. (1998). Model-based geostatistics. *Applied Statistics, 47*, 229–350.

Dlamini, N., Zulu, Z., Kunene, S., Geoffroy, E., Ntshalintshali, N., Owiti, P., Sikhondze, W., Makadzange, K., & Zachariah, R. (2018). From diagnosis to case investigation for malaria elimination in Swaziland: Is reporting and response timely? *Public Health Action, 8*, S8–S12.

Erbach-Schoenberg, E. Z., Alegana, V. A., Sorichetta, A., Linard, C., Lourenço, C., Ruktanonchai, N. W., Graupe, B., Bird, T. J., Pezzulo, C., Wesolowski, A., et al. (2016). Dynamic denominators: The impact of seasonally varying population numbers on disease incidence estimates. *Population Health Metrics, 14*, 35.

Fahrmeir, L., & Knorr-Held, L. (2000). Dynamic and semiparametric models. In *Smoothing and regression*. New York: John Wiley & Sons.

Gemperli, A., Vounatsou, P., Sogoba, N. & Smith, T. (2006). Malaria Mapping Using Transmission Models: Application to Survey Data from Mali. American Journal of Epidemiology, *163*, 289–297.

Giorgi, E., Sesay, S. S. S., Terlouw, D. J., & Diggle, P. J. (2015). Combining data from multiple spatially referenced prevalence surveys using generalized linear geostatistical models. *Journal of the Royal Statistical Society: Series A (Statistics in Society), 178*, 445–464.

Giorgi, E., Osman, A. A., Hassan, A. H., Ali, A. A., Ibrahim, F., Amran, J. G. H., Noor, A. M., & Snow, R. W. (2018). Using non-exceedance probabilities of policy-relevant malaria prevalence thresholds to identify areas of low transmission in Somalia. *Malaria Journal, 17*, 88.

Githinji, S., Oyando, R., Malinga, J., Ejersa, W., Soti, D., Rono, J., Snow, R. W., Buff, A. M., & Noor, A. M. (2017). Completeness of malaria indicator data reporting via the District Health Information Software 2 in Kenya, 2011–2015. *Malaria Journal, 16*, 344.

Guerra, C., Hay, S., Lucioparedes, L., Gikandi, P., Tatem, A., Noor, A., & Snow, R. (2007). Assembling a global database of malaria parasite prevalence for the Malaria Atlas Project. *Malaria Journal, 6*, 17.

Guerra, C. A., Gikandi, P. W., Tatem, A. J., Noor, A. M., Smith, D. L., Hay, S. I., & Snow, R. W. (2008). The limits and intensity of Plasmodium falciparum transmission: Implications for malaria control and elimination worldwide. *PLoS Medicine, 5*, e38.

Hay, S. I., & Snow, R. W. (2006). The malaria Atlas Project: Developing global maps of malaria risk. *PLoS Medicine, 3*, e473.

Hay, S. I., Guerra, C. A., Tatem, A. J., Noor, A. M., & Snow, R. W. (2004). The global distribution and population at risk of malaria: Past, present, and future. *The Lancet Infectious Diseases, 4*, 327–336.

Hay, S. I., Smith, D. L., & Snow, R. W. (2008). Measuring malaria endemicity from intense to interrupted transmission. *The Lancet Infectious Diseases, 8*, 369–378.

Hay, S. I., Battle, K. E., Pigott, D. M., Smith, D. L., Moyes, C. L., Bhatt, S., Brownstein, J. S., Collier, N., Myers, M. F., George, D. B. & Gething, P. W. (2013). Global mapping of infectious disease. Philosophical Transactions of the Royal Society B: Biological Sciences, 368.

Hsiang, M. S., Hwang, J., Kunene, S., Drakeley, C., Kandula, D., Novotny, J., Parizo, J., Jensen, T., Tong, M., Kemere, J., et al. (2012). Surveillance for malaria elimination in Swaziland: A national cross-sectional study using pooled PCR and serology. *PLoS One, 7*, e29550.

Huete, A., Didan, K., Miura, T., Rodriguez, E. P., Gao, X. & Ferreira, L. G. (2002). Overview of the radiometric and biophysical performance of the MODIS vegetation indices. Remote Sensing of Environment, *83*, 195–213.

Huffman, G. J., Bolvin, D. T., Nelkin, E. J., Wolff, D. B., Adler, R. F., Gu, G., Hong, Y., Bowman, K. P. & Stocker, E. F. (2007). The TRMM Multisatellite Precipitation Analysis (TMPA): Quasi-Global, Multiyear, Combined-Sensor Precipitation Estimates at Fine Scales. *Journal of Hydrometeorology, 8*, 38–55.

Joncour, G. (1956). La lutte contre le paludisme a Madagascar. *Bulletin of the World Health Organization* (pp. 711–723).

Karuri, J., Waiganjo, P., & Manya, A. (2014). DHIS2: The tool to improve health data demand and use in Kenya. *Journal of Health Informatics in Developing Countries, 8*. Retrieved from https://www.jhidc.org/index.php/jhidc/article/view/113.

Kazembe, L. N. (2007). Spatial modelling and risk factors of malaria incidence in northern Malawi. Acta Tropica, *102*, 126–137.

Lariviere, M., Hocquet, P., & Abonnenc, E. (1961). Résultats d'une enquête palustre dans la république du Sénégal. Indices plasmodiques chez les enfants en milieu rural. *Bulletin de la Société Médicale d'Afrique Noire de Langue Française* (pp. 386–403).

Le Sueur, D., Binka, F., Lengeler, C., De Savigny, D., Snow, B., Teuscher, T., & Toure, Y. (1997). An atlas of malaria in Africa. *Africa Health, 19*, 23–24.

Macdonald, G. (1950). The analysis of malaria parasite rates in infants. *Tropical Diseases Bulletin, 47*, 915–938.

Macdonald, G. (1957). *The epidemiology and control of malaria*. London: Oxford University Press.

Maina, J., Ouma, P. O., Macharia, P. M., Alegana, V. A., Mitto, B., Fall, I. S., Noor, A. M., Snow, R. W., & Okiro, E. A. (2019). A spatial database of health facilities managed by the public health sector in sub Saharan Africa. *Scientific Data, 6*, 134.

Mccrae, A. (Ed.). (1975). *Uganda atlas of disease distribution*. Nairobi: East African Publishing House.

Molineaux, L., Muir, D. A., Spencer, H. C., & Wernsdorfer, W. H. (1988). The epidemiology of malaria and its measurement. In W. H. Wernsdorfer & I. Mcgregor (Eds.), *Malaria: Principles and practice of malariology*. London: Churchill Livingstone.

Mueller, I., Galinski, M. R., Baird, J. K., Carlton, J. M., Kochar, D. K., Alonso, P. L., & Del Portillo, H. A. (2009). Key gaps in the knowledge of Plasmodium vivax, a neglected human malaria parasite. *The Lancet Infectious Diseases, 9*, 555–566.

Mueller, I., Slutsker, L., & Tanner, M. (2011). Estimating the burden of malaria: The need for improved surveillance. *PLoS Medicine, 8*, e1001144.

Murtaugh, P. A. (2009). Performance of several variable-selection methods applied to real ecological data. *Ecology Letters, 12*, 1061–1068.

Noor, A. M., Amin, A. A., Gething, P. W., Atkinson, P. M., Hay, S. I., & Snow, R. W. (2006). Modelling distances travelled to government health services in Kenya. *Tropical Medicine & International Health, 11*, 188–196.

Noor, A., Clements, A., Gething, P., Moloney, G., Borle, M., Shewchuk, T., Hay, S., & Snow, R. (2008). Spatial prediction of Plasmodium falciparum prevalence in Somalia. *Malaria Journal, 7*, 159.

Noor, A. M., Gething, P. W., Alegana, V. A., Patil, A. P., Hay, S. I., Muchiri, E., Juma, E., & Snow, R. W. (2009). The risks of malaria infection in Kenya in 2009. *BMC Infectious Diseases, 9*, 180.

Noor, A. M., Alegana, V. A., Patil, A. P., Moloney, G., Borle, M., Yusuf, F., Amran, J., & Snow, R. W. (2012). Mapping the receptivity of malaria risk to plan the future of control in Somalia. *BMJ Open, 2*, e001160.

Noor, A., Uusiku, P., Kamwi, R., Katokele, S., Ntomwa, B., Alegana, V., & Snow, R. (2013a). The receptive versus current risks of Plasmodium falciparum transmission in northern Namibia: Implications for elimination. *BMC Infectious Diseases, 13*, 184.

Noor, A. M., Alegana, V. A., Kamwi, R. N., Hansford, C. F., Ntomwa, B., Katokele, S., & Snow, R. W. (2013b). Malaria control and the intensity of *Plasmodium falciparum* transmission in Namibia 1969–1992. *PLoS One, 8*, e63350.

Noor, A. M., Kinyoki, D. K., Mundia, C. W., Kabaria, C. W., Mutua, J. W., Alegana, V. A., Fall, I. S., & Snow, R. W. (2014). The changing risk of Plasmodium falciparum malaria infection

in Africa: 2000-10: A spatial and temporal analysis of transmission intensity. *The Lancet, 383*, 1739–1747.
Pull, J. H. (1972). Malaria surveillance methods, their uses and limitations. *The American Journal of Tropical Medicine and Hygiene, 21*, 651–657.
Ray, A. P., & Beljaev, A. E. (1984). Epidemiological surveillance: A tool for assessment of malaria and its control. *The Journal of Communicable Diseases, 16*, 197–207.
Schrödle, B., & Held, L. (2010). A primer on disease mapping and ecological regression using INLA. *Computational Statistics, 26*, 241–258.
Schrödle, B., & Held, L. (2011). Spatio-temporal disease mapping using INLA. *Environmetrics, 22*, 725–734.
Simmons, J. S., Whayne, T. F., Anderson, G. W., Division, U. S. S.-G. S. O. P. M., Service, U. S. S.-G. S. O. P. M. & Horack, H. M. 1951. Global Epidemiology: A Geography of Disease and Sanitation, J.B. Lippincott.
Snow, R. (2014). Sixty years trying to define the malaria burden in Africa: Have we made any progress? *BMC Medicine, 12*, 227.
Snow, R., Marsh, K., & Lesueur, D. (1996). The need for maps of transmission intensity to guide malaria control in Africa. *Parasitology Today, 12*, 455–457.
Snow, R. W., Guerra, C. A., Noor, A. M., Myint, H. Y., & Hay, S. I. (2005). The global distribution of clinical episodes of Plasmodium falciparum malaria. *Nature, 434*, 214–217.
Stekelenburg, J., Jager, B. E., Kolk, P. R., Westen, E. H. M. N., Kwaak, A. V. D., & Wolffers, I. N. (2005). Health care seeking behaviour and utilisation of traditional healers in Kalabo, Zambia. *Health Policy, 71*, 67–81.
Sturrock, H. J., Novotny, J. M., Kunene, S., Dlamini, S., Zulu, Z., Cohen, J. M., Hsiang, M. S., Greenhouse, B., & Gosling, R. D. (2013). Reactive case detection for malaria elimination: Real-life experience from an ongoing program in Swaziland. *PLoS One, 8*, e63830.
Swaroop, S., Gilroy, A. B., & Uemura, K. (1966). Statistical methods in malaria eradication. *Monograph Series. World Health Organization, 51*, 1–164.
Tatem, A. J. (2014). Mapping the denominator: Spatial demography in the measurement of progress. *International Health, 6*, 153–155.
Twohig, K. A., Pfeffer, D. A., Baird, J. K., Price, R. N., Zimmerman, P. A., Hay, S. I., Gething, P. W., Battle, K. E., & Howes, R. E. (2019). Growing evidence of Plasmodium vivax across malaria-endemic Africa. *PLoS Neglected Tropical Diseases, 13*, e0007140.
White, N. (2011). Determinants of relapse periodicity in Plasmodium vivax malaria. *Malaria Journal, 10*, 297.
WHO. (2015). *Global technical strategy for malaria 2016–2030*. Geneva: WHO.
World Health Organization. (2012a). *Disease surveillance for malaria control: An operational manual*. Geneva: World Health Organization.
World Health Organization. (2012b). *Test. Treat. Track. Scaling up diagnostic testing, treatment and surveillance for malaria*. Geneva: World Health Organization.

Chapter 4
Geographically Precise Public Health: Case of Southern Mozambique

Newton Nyapwere, Yolisa Dube, and Prestige Tatenda Makanga

Abstract The spatial turns in health and other fields have opened new ways of data collection and analysis to reveal new insights in health sciences. Geographical information science has much potential of being applied in precision public health for community-level interventions. The objective of this chapter is to demonstrate how the geospatial sciences are increasingly an important aspect of this new approach public health. The chapter will first look at the principles of precision public health and how the transition from precision medicine to precision public health has evolved. A case study on spatial epidemiology in maternal health in Mozambique is introduced to demonstrate the value brought by geography to precision public health. Lastly, a discussion on the geographically enabled precision public health focuses on future challenges and opportunities.

Keywords Geographical information science · Precision public health · Maternal outcomes · Precision medicine · Precision prevention

Introduction

The geographical information science (GISc) is playing an enormous role in every field of endeavour with public health not spared. The huge explosion in GISc in health and other fields has opened new avenues for data collection and analysis to reveal new insights in public health data that have previously been inapparent. With the dramatic improvements in database technology in many ways analogous to the driving forces, current advances in data generation and handling in biomedicine, it became apparent to many users of geographically indexed information that a surprisingly high portion of the world's information could be organised around coordinates (Bardakjian and Gonzalez-Alegre 2018). Thus, organising data in space and

N. Nyapwere (✉) · Y. Dube · P. T. Makanga
Place Alert Labs (PALs), Department of Surveying and Geomatics,
Midlands State University, Gweru, Zimbabwe
e-mail: nyapweren@staff.msu.ac.zw

P. T. Makanga (ed.), *Practicing Health Geography*, Global Perspectives on Health Geography, https://doi.org/10.1007/978-3-030-63471-1_4

time has contributed to the exposure of trends in data that were somewhat hidden before.

Precision public health is defined as the application and combination of new and existing technologies, which more precisely describe and analyse individuals and their environment over the life course, in order to tailor preventive interventions for at-risk groups and improve the overall health of the population (Dolley 2018). Geospatial approaches have emerged to be powerful tools in precision public health. This is attributed to the fact that precision public health includes the need to have knowledge of time, space and its socioeconomic and physical characteristics associated with the place. Thus, geography plays a role in organising population health data to find patterns and behaviour that increase knowledge of precision public health and maintain a healthy population through prevention and monitoring. Geographical Information Systems (GIS) include the processes, infrastructure, software and the people involved in coming up with location-based solutions for improved public health outcomes. On the other hand, GISc includes the use of theories, methods, technology, and data for understanding geographic processes, relationships, and patterns (Goodchild 2010).

Precision Public Health

Before precision public health was coined in 2013 (Gudes et al. 2018), much of the work was focused on precision medicine. Due to precision medicine's focus on treatment rather than prevention, it is unlikely to improve population health (Khoury and Galea 2016). Thus, focusing more on trying to prevent disease before there is an outbreak should be priority as this can reduce costs and loss of lives associated with an outbreak of disease. Precision medicine can be described as treatment strategies that take individual clinical variability into account (Bardakjian and Gonzalez-Alegre 2018). In other words, it is an approach for disease treatment that considers individual variability in genes, environment and socioeconomic status. This paradigm relies on the narrative that an ever-increasing knowledge of biological mechanisms, particularly genetics, coupled with information technology leads to transformative improvements in population health (Rey-López et al. 2018). Precision medicine is needed because a high percentage of patients using a particular drug class may derive no benefit from the drug (Spear et al. 2001) due to variability in physiological makeup. Therefore, it is important that individuals are treated according to their uniqueness, both genetically and phenotypically (Weeramanthri and Woodgate 2016). Precision medicine has enabled medical doctors to use patients' genetic and other molecular information as part of routine delivery of medical care (Bilkey et al. 2019). It also enables improved ability to predict which treatments will work best for specific health conditions.

Although precision medicine focuses on individualised care, it is hard to infer causality by working from individual observations (Khoury and Evans 2015). Precision medicine has limitations which compelled public health researchers and medical practitioners to shift focus to a more robust intervention approach. Bayer and Glalea (2015) have noted that health differences among different social groups are less likely driven by better clinical treatments but by marked differences in socioeconomic factors. These social factors could be used to stratify populations for public health interventions. Could the same technologies that propel precision medicine usher in a parallel era of precision public health beyond treatment of sick individuals (Khoury et al. 2016)? This is still yet to be realised. Precision public health rides on the success and principles of precision medicine in that it looks at individual-level data and aggregates that information at population level (Bilkey et al. 2019). Characterising and describing a disease outcome in a subset of populations helps in rapid prevention and treatment. The approach is less costly and can be used for policy and planning at a community, regional or national scale. This contrasts with simply concentrating treatment of individuals without a scaling up approach at levels with similar characteristics and disease outcomes. Precision public health main promise is on the potential for prevention rather than treatment.

Geography and Precision Public Health

Geography in the physical sense of the word refers to the spatio-temporal characteristics contained in a given area. In other instances, the term is used to describe the contextual, socioeconomic and cultural dynamics at a given place. Knowing the characteristics and distribution of social and environmental determinants in a given place in time helps to put in place preventive measures such as targeted awareness campaigns, immunisations, location-based early warning systems and formulating and targeting favourable public health policies and programmes. Exposing risk geographically helps to visualise where it is greatest and thus applying specific prevention measures unique to that area at the right time. GIS-based predictive modelling offers a useful avenue for guiding public health policy and obtaining high precision data on population risks, comorbidities, changes in incidence rates, effectiveness of treatment programmes and socioeconomic factors associated with risk (Fleisch et al. 2015; MacQuillan et al. 2017; Yazdy et al. 2015).

The use of geographic data in health is not new. While cartography was performed before by geographers to create maps, the most notable individual to use mapping techniques in health was the English Physician, John Snow. Regarded as the father of modern day epidemiology, John Snow created maps of cholera-related deaths in London during the most severe 1854 epidemic (Bavley et al. 2013). In addition, these maps were also used as the basis for targeting the appropriate health intervention in the form of placing a lock on the water pump that was the main source of cholera in the affected communities. This resulted in an immediate decline of cholera cases and deaths.

The bulk of geo-coded health data are collected at health facilities. However, in certain circumstances, there is need to collect health data from the population such as at the place of residence. Various geospatial data collection methods in the field have evolved with the proliferation the GPS, smartphones and the associated IT products. Data on individuals in a population such as pregnant women, disease characteristics and health facilities and access to health care can now easily be gathered or obtained in the field and on online platforms using different approaches such as satellite imagery, databases, GPS devices, smartphones and mobile applications (e.g. Open Data Kit), participatory GIS and more recently drones.

Framework spatial data on roads, rivers and community boundaries can be obtained through use of high-resolution open source data imagery such as Google Earth, Open Street Map (OSM) and Bing map as well using the participatory GIS techniques. GPS and smartphone devices can be used during household surveys, tracking patients for research purposes, as well as for follow up for treatment and prevention of diseases. GPS devices are used to capture coordinates for health facilities, households and disease incidence during health mapping exercises. The data collected can be used in spatial epidemiology, using geography as an organising principle for understanding associations between determinants of health and health outcomes and how these vary geographically.

Understanding the natural history of a disease, history of patients and routine checkups plays a pivotal role in understanding the causal pathways of a disease (Newnham et al. 2017). To say we have fully implemented precise monitoring strategies the question is not only "What health matter has affected Who?" but also "Where and When?" The last two questions are answered by applying the four dimensions of geographical space, the ground coordinates and time. Early diagnosis and intervention are key to reducing disease burden (Dolley 2018).

Precision public health relies on linking population data to measurable outcomes and stratifying it according to place, time and persons (Khoury and Galea 2016). Focus is on geospatial methods and data that are used to monitor, and hence detect and stratify risk according to place, time and persons to predict place specific future outbreaks. This chapter illustrates these principles and concepts as applied to maternal health in Mozambique.

Place Specific Factors Related to Maternal Health in Mozambique

The aim of this study was to identify and measure the community specific determinants that were associated with maternal and perinatal ill health in the southern region of Mozambique. In line with the recommendations from the Sustainable Development Goals (SDGs), the study sought to gain a local understanding of these determinants and how their associations with adverse maternal outcomes varied geographically.

Data Collection

The set of context-specific variables were derived from a scoping review, and community consultation through focus group discussions and semi-structured interviews. Individual woman level data measuring these variables were collected through a household census conducted as part of the Community Level Intervention for Preeclampsia (CLIP) feasibility study. The census included information on all women who had been pregnant in the 12 months prior to the census, as well as women of reproductive age who had died. Data collected included individual-level variables (e.g. age, education and pregnancy history), as well as community characteristics (e.g. availability of the household head and community support initiatives). All reports of maternal, foetal or perinatal fatality were followed up with verbal autopsy to classify the cause of death.

The census identified 50,493 households that were home to 80,483 women of reproductive age (age 12–49 years). Of these women, 14,617 had been pregnant in the 12 months prior to the census, of which 9172 (61.6%) had completed their pregnancies. For the mother, there were 18 deaths (204.6 MMR) of which the verbal autopsy identified that 38% were from direct causes and 62% from indirect causes. For the baby, there were 288 (3.0%) miscarriages, 466 (4.9%) stillbirths and 8796 (92.1%) live births, of which there were 117 neonatal deaths. A full description of the health and sociodemographic profile of the women of reproductive age in the study area has been published.

In addition, the study collected five geographical variables using geographical information systems. There were three travel times to (1) primary health facilities, (2) secondary health facilities and (3) tertiary health facilities, using mixed transport modes for public transport and ambulances. Walking times to the nearest main road (4) were calculated to measure the degree to which communities were geographically isolated. Finally, an indicator for flood proneness (5) was designed based on flood and precipitation records from the previous year. These variables and other community-level estimates for the variables captured in the census were calculated for each locality (administrative unit) in the study area as described in Table 4.1. Both the census and geographical data were aggregated into community-level averages at the locality level for each of the chosen variables, as ethical approval did not allow to analyse the location data at the level of the individual woman.

Statistical Methods

The primary outcome for this study was a combined maternal and perinatal outcome that included maternal, foetal and neonatal deaths. The denominator was the total number of live births. A composite outcome was chosen as powering the study for maternal deaths alone would have required a prohibitively large sample size. There is clinical plausibility in combining the three outcomes as both foetal and early

Table 4.1 Community-level variable considered in this study

Community-level variable	Description (*Variables calculated for reproductive age women with completed pregnancies*)
Census variables	
1. Age of reproductive age woman	Average age of reproductive age woman
2. Household head's education	Average number of years that household heads (man or woman) have spent in school (*No schooling = 0; at least primary = 7; at least secondary = 12; at least a degree = 16; graduate = 18, post grad = 20*)
3. Household head's availability	Percentage of households where the household head lives in the house
4. Water source score	Percentage of households that have an improved water source
5. Latrine score	Percentage of households that have an improved latrine
6. Private transportation score	Percentage of reproductive age women who live in a house where someone owns a private car
7. Reproductive age women's education	Average number of years that reproductive age women have spent in school (*No schooling = 0, grade 5 = 5, grade 7 = 7, grade 10 = 10, grade 12 = 12, bachelors = 16, graduate = 18, post grad = 20*)
8. Fertility rate	Average number of children born to each woman in the community that had a completed pregnancy
9. Reproductive age women's marital status score	Percentage of reproductive age women in a marital union (monogamous or polygamous) relative to total with completed pregnancies
10. Reproductive age women's unemployment rate	Proportion of reproductive age women that do not work compared with total reproductive age women with a completed pregnancy
11. Family support	Percentage of reproductive age women that would receive financial, transport and emotional help from family or neighbours for a pregnancy related need
12. Community group support	Percentage of reproductive age women that would receive financial, transport and emotional help from a community-based group for a pregnancy related need
13. Financial autonomy in pregnancy	Percentage households where the reproductive age woman is empowered to make financial decisions concerning her pregnancy
Geospatial variables	
14. Access to primary health facilities	Average travel time to the nearest primary health facility, using public transport
15. Access to secondary health facilities	Average travel time to the nearest secondary health facility, using a mix of public transport and an ambulance
16. Access to tertiary health facilities	Average travel time to the nearest tertiary health facility, using a mix of public transport and an ambulance
17. Isolation	Average walking time to the nearest main road
18. Flood proneness	The difference between the road quality indicator (RoQI) score on a typical day in the dry season and on the worst day in the wet season. RoQI scores range between 0 and 100 and are a function of the quality of roads in a community

neonatal outcomes are related to the woman's condition during the antenatal and intrapartum periods, while her environment and sociocultural circumstances have an impact on late neonatal outcomes. Furthermore, a significant proportion of miscarriages are related to placental dysfunction, as are stillbirths and neonatal deaths.

The spatial statistics module within ArcGIS software was used for exploratory regression to further prioritise variables and to create the global ordinary least squares (OLS) regression model. The exploratory regression exercise evaluated different combinations of our explanatory variables for their fit for an OLS model and how these explained trends in our outcome variable. This method implements the exploration by screening variables in a forward stepwise sequence, exploring how different combinations of variables fit and perform in the regression model.

The geographically weighted regression (GWR) technique was used to develop a second model, which extended the output from OLS, to explore spatial non-stationarity effect of the variables. This allowed for the new model to account for spatial structure in estimating local rather than global model parameters. We foresee this to be an important step to creating interventions that are locally specific and an important part of more precisely targeting interventions. As part of the modelling process, the spatial weights based on the geographic proximity of observation are applied to give more weight to values that were closer together.

Results

A total of six variables (Table 4.2) were statistically significant ($P \leq 0.05$) in explaining the combined maternal and perinatal outcome. These included: geographic isolation, flood proneness, access to an improved latrine, average age of reproductive age woman, family support and fertility rates. The performance of the ordinary least squares model was an adjusted $R^2 = 0.69$. Three of the variables (isolation, latrine score and family support) showed significant geographic variability in their effect on rates of adverse outcome. Accounting for this modest non-stationary

Table 4.2 OLS model

Variable	Coefficient	SE	*T*-statistic	VIF
Intercept	−0.194483	0.19802	−0.98214	–
Isolation	0.033353*	0.01464	2.27805	2.06598
Flood proneness (%)	0.023936**	0.00754	3.17587	1.35311
Latrine support (%)	0.003094***	0.00072	−4.28457	1.92347
Family support (%)	0.001274*	0.00051	−2.48055	1.52648
Age of reproductive woman age (years)	0.034653***	0.00826	4.19544	2.93335
Fertility rate (number of children)	−0.222607***	0.02942	−7.56575	3.90853

Multiple $R^2 = 0.75$; adj $R^2 = 0.69$. *OLS* ordinary least squares, *VIF* variance inflation factor
*$P \leq 0.05$; **$P \leq 0.01$; ***$P \leq 0.001$

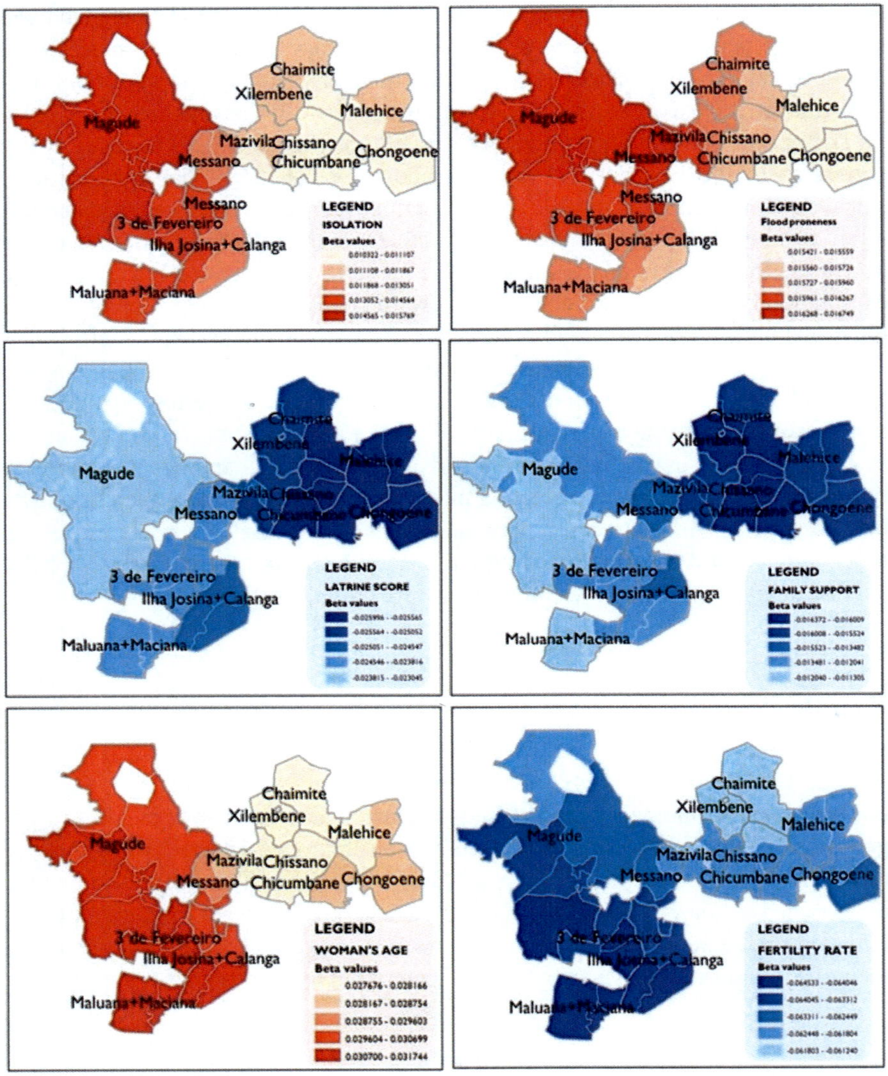

Fig. 4.1 Geographic variation in the effect of the combined outcome using beta coefficients. Darkening shade of blue indicates increasing protective effect while darkening shade of red indicates increasing risk effect

effect through geographically weighted regression increased the adjusted R^2 to 0.71 (Fig. 4.1).

The GWR model indicated that the effect on the combined outcome was geographically non-stationary for three of the six variables (isolation, latrine score and family support). There was no non-stationarity for the effect of flood proneness, age of the woman and fertility rate. The performance of the local model improved modestly from the global model to an adjusted R^2 value of 0.71, explaining a further 2%

of the variability in the outcome. A graduated colour classification was used to describe the magnitude of the effect in each of the localities.

The general direction of the effects in the global model is preserved in the local model. Isolation has a general effect of increasing risk for adverse outcomes across the entire study area. However, the effect is more pronounced by a factor of approximately 53% in the western region of Magude compared with the east in Chongoene. A similar pattern is observed for the effect of flood proneness on the combined outcome, with the western regions of the study area being at greater risk, though the magnitude of the differential effect is much smaller (8%). This is also even though the same areas in the west had lower level of proneness to floods.

The proportion of household with an improved latrine was associated with decreasing the adverse outcomes for all regions under study, although this effect is greatest in magnitude for the eastern region of the study area by a factor of about 12%. Family support has a greater effect of reducing the rates of adverse outcomes in the eastern regions by a factor of about 45%. Average woman's age had an effect of increasing the outcome by up to 15% more in the west than the east. The variability in the association of fertility rates to the rates of adverse outcomes was also non-stationary with the highest effect 5% more than the lowest.

Discussion

The case study shows how geographical methods help to better understand the place specific linkages between determinants of maternal health and associated outcomes. This is an important aspect for designing relevant location specific precision public health policies and programmes. This will help in precision prevention and monitoring which works on the principle of targeted intervention. The evidence produced in this case study highlights how the effect of different variables on health outcomes differs from one place to the next. This knowledge is critical for achieving impact and improving health in a resource limited setting like Mozambique.

One of the key challenges that remains concerns the translation of much of this evidence on the place specific character of associations into decision aids and tools that precisely target interventions. This is true both for policy and programme formulation. In terms of policy, the evidence we have suggests that it may be futile to create blanketing national level policies without taking into account subnational patterns in disease outcomes and associated determinants. New policy directives must be informed by the granular patterns in both. With regards to programmes for intervention, our study points to a need for place sensitivity in order to have the greatest impact. There is much potential for translating evidence such as that produced in this study into GPS enabled mobile health tools that that provide near real time place specific strategies to help public health personnel on the ground to make more informed decisions (see Chap. 14 for example). These decisions naturally go beyond what's clinical, by addressing the broader socio-geographical characteristic that shape health outcomes.

While this project leveraged an existing highly resourced project to produce good quality GIS data that were finely granulated geographically, framework GIS data for health-related research in the Africa remains scarce and expensive to generate. Data sharing efforts are still poorly coordinated and there is a lot of duplication in data creation efforts. Even when the data have been made available to others, interoperability is sometimes a challenge because of the different data formats for archiving and dissemination that are used by different organisations. GIS data should have the ability to be exchanged, integrated and readily used by other users without undergoing rigorous data conversions which may compromise its quality. Therefore, there is a need for a home grown African or country specific spatial data repositories for public health data and information which should be shared among organisation and countries in Africa as open source or as proprietary data at a finer resolution. Consequently, there is a need for cooperation between academia, government and non-governmental organisations in public health to work towards improved data sharing and integration through the development of data standards and policies such as information sharing networks.

Conclusion

Precision public health is a relatively new and rapidly evolving field of health. The application of geographical thought and methods helps to better specify trends in disease occurrence and the associated determinants. Adding a geographic perspective to precision public health will tremendously increase as has already occurred in the broader public health, population studies and medical fields. Mobile technologies are rapidly penetrating in African populations and it is against this background that GISc can ride on the mobile technology to easily and readily collect relevant data from individuals that is useful for precision public health.

References

Bardakjian, T., & Gonzalez-Alegre, P. (2018). Towards precision medicine. In *Handbook of clinical neurology* (Vol. 147). Amsterdam: Elsevier. https://doi.org/10.1016/B978-0-444-63233-3.00008-7.

Bavley, R., Musa, G. J., Tsou, H., Sylk, T., Keating, W., Lakew, B., et al. (2013). Use of GIS mapping as a public health tool—From cholera to cancer. *Health Services Insights, 6*, 111–116. https://doi.org/10.4137/hsi.s10471.

Bayer, R., & Glalea, S. (2015). Public health in precision medicine era. *The New England Journal of Medicine, 70*(12), 741–742. https://doi.org/10.1097/01.ogx.0000473688.62559.e2.

Bilkey, G. A., Burns, B. L., Coles, E. P., Mahede, T., Baynam, G., & Nowak, K. J. (2019). Optimizing precision medicine for public health. *Frontiers in Public Health, 7*(March), 1–9. https://doi.org/10.3389/fpubh.2019.00042.

Dolley, S. (2018). Big data's role in precision public health. *Frontiers in Public Health, 6*, 12. https://doi.org/10.3389/fpubh.2018.00068.

Fleisch, A. F., Rifas-Shiman, S. L., Koutrakis, P., Schwartz, J. D., Kloog, I., Melly, S., et al. (2015). Prenatal exposure to traffic pollution: Associations with reduced fetal growth and rapid infant weight gain. *Epidemiology, 26*(1), 43–50. https://doi.org/10.1097/EDE.0000000000000203.

Goodchild, M. F. (2010). Twenty years of progress: GIScience in 2010. *Journal of Spatial Information Science, 1*(1), 3–20. https://doi.org/10.5311/josis.2010.1.2.

Gudes, O., Weeramanthri, T. S., Baynam, G., Dawkins, H. J. S., Semmens, J. B., & Bellgard, M. (2018). Editorial: Precision public health. *Frontiers in Public Health, 6*(April), 3–5. https://doi.org/10.3389/fpubh.2018.00121.

Khoury, M. J., & Evans, J. P. (2015). A public health perspective on a national precision medicine cohort. *HHS Public Access, 313*(21), 2117–2118. https://doi.org/10.1001/jama.2015.3382.A.

Khoury, M. J., & Galea, S. (2016). Will precision medicine improve population health? *Obstetrical and Gynecological Survey, 72*(1), 15–17. https://doi.org/10.1097/01.ogx.0000511937.49229.57.

Khoury, M. J., Iademarco, M. F., Riley, W. T., Sciences, P., Service, P. H., Corps, C., et al. (2016). Precision public health for the era of precision medicine. *American Journal of Preventive Medicine, 50*(3), 398–401. https://doi.org/10.1016/j.amepre.2015.08.031.Precision.

MacQuillan, E. L., Curtis, A. B., Baker, K. M., Paul, R., & Back, Y. O. (2017). Using GIS mapping to target public health interventions: Examining birth outcomes across GIS techniques. *Journal of Community Health, 42*(4), 633–638. https://doi.org/10.1007/s10900-016-0298-z.

Makanga, P. T., Sacoor, C., Schuurman, N., Lee, T., Vilanculo, F. C., Munguambe, K., et al. (2019). Place-specific factors associated with adverse maternal and perinatal outcomes in Southern Mozambique: A retrospective cohort study. *BMJ Open, 9*(2), e024042. https://doi.org/10.1136/bmjopen-2018-024042.

Newnham, J. P., Arrese, C. A., Kemp, M. W., Hart, R. J., Keelan, J. A., & White, S. W. (2017). Applying precision public health to prevent preterm birth. *Frontiers in Public Health, 5*(April), 1–16. https://doi.org/10.3389/fpubh.2017.00066.

Rey-López, J. P., de Sá, T. H., & de Rezende, L. F. M. (2018). Why precision medicine is not the best route to a healthier world. *Revista de Saude Publica*. https://doi.org/10.11606/S1518-8787.2018052000209.

Spear, B. B., Heath-Chiozzi, M., & Huff, J. (2001). Clinical application of pharmacogenetics. *Trends in Molecular Medicine, 7*(5), 201–204. https://doi.org/10.1016/S1471-4914(01)01986-4.

Weeramanthri, T. S., & Woodgate, P. (2016). Spatially enabling the health sector. *Frontiers in Public Health, 4*(November), 6–10. https://doi.org/10.3389/fpubh.2016.00243.

Yazdy, M. M., Werler, M. M., Feldkamp, M. L., Shaw, G. M., Mosley, B. S., & Vieira, V. M. (2015). Spatial analysis of gastroschisis in the national birth defects prevention study. *Birth Defects Research Part A—Clinical and Molecular Teratology, 103*(6), 544–553. https://doi.org/10.1002/bdra.23375.

Chapter 5
Spatial Epidemiology of Urban Health Risks in Select West African Cities

Ibrahima Sy, Sokhna Thiam, Richard Mbra Kouassi, Doulo Traoré, Birane Cissé, Brama Koné, Jacques-André Ndione, Peter Steinmann, Aminata Niang Diène, Badara Cissé, Juerg Utzinger, Guéladio Cissé, and Marcel Tanner

Abstract West African cities face critical societal challenges that are linked to environmental and health changes. These challenges are further exacerbated by urbanization dynamics, climate change, socio-economic mutation and lack of capacity for sustainable urban development, governance and basic services delivery. The deficiency of environmental sanitation and ecosystem services have led to high complexity of urban health risks inequalities, resulting in the need for more research on efficient urban health policies. The purpose of this contribution is to present the main findings on the spatial epidemiology of diarrhaea and malaria, and their associated risks factors in the following select West African cities. Spatial variability of exposure to diarrhaea and malaria transmission is linked to several health risks such as lack of access to water and sanitation, solid wastes management, urban flooding,

I. Sy (✉)
Université Cheikh Anta Diop (UCAD), Dakar, Senegal

Centre de Suivi Ecologique (CSE), Dakar, Senegal

Swiss Tropical and Public Health Institute (Swiss TPH), Basel, Switzerland
e-mail: ibrahima41.sy@ucad.edu.sn

S. Thiam · P. Steinmann · J. Utzinger · G. Cissé · M. Tanner
Swiss Tropical and Public Health Institute (Swiss TPH), Basel, Switzerland

R. M. Kouassi · B. Koné
Centre Suisse de Recherches Scientifiques en Côte d'Ivoire (CSRS), Abidjan, Côte d'Ivoire

D. Traoré
Action Contre la Faim-Espagne (ACF-E), Nouakchott, Mauritania

B. Cissé · J.-A. Ndione
Centre de Suivi Ecologique (CSE), Dakar, Senegal

A. N. Diène
Université Cheikh Anta Diop (UCAD), Dakar, Senegal

B. Cissé
Institut de Recherche en Santé, de Surveillance Epidémiologique et de Formation (IRESSEF), Dakar, Senegal

P. T. Makanga (ed.), *Practicing Health Geography*, Global Perspectives on Health Geography, https://doi.org/10.1007/978-3-030-63471-1_5

precarious housing conditions, urban agriculture, socio-economic status and social practices and individual behaviours fluctuating according to the specificity of urban areas. These factors and their interconnections make some groups (differentiated by geography) more vulnerable to poor outcomes due to limited capacity to effectively respond to health risks and poor performance by the health system. We demonstrate the potential for a spatial epidemiology perspective that disaggregates data to, clusters of houses, neighbourhood, sector and municipality scale to support health policies and interventions, and how these results can contribute to achieving SDG 3 (Health and Well-Being for all) in African urban areas.

Keywords Spatial epidemiology · Diseases · Risks factors · Public health · Cities · West Africa

Introduction

Complexity of Urban Health Challenges in West African Cities

Several scientific studies have demonstrated that urbanization is integrally connected to the three pillars of sustainable development: economic development, social development and environmental protection (WHO-UN Habitat 2010). The process of urban population growth generates important public health challenges for city planners and governments (Szombathely et al. 2017). With the demographic and epidemiological transition (people living longer, change in population health due to changes in lifestyle), urbanization poses major public health risk as urban populations are rapidly increasing, while the basic infrastructure remains insufficient. Social and economic inequities in urban areas have also resulted in significant health inequalities (Salem 1998; Kjellstrom et al. 2007). However, when cities are well managed, they offer important opportunities by facilitating urban integration, social development and the improvements of population health with the increase of health care access (Sy 2006).

West African cities pose particular challenges to the health sustainable development goal. Some factors that have an impact on human health include: (1) unplanned settlements (State of African cities, 2014), (2) rapid emergence of informal economies (IIED 2016) and (3) new lifestyle linked to poverty and social change and intense pressure on natural resources. In the context of climate change, it is much harder to provide services such as fresh water, waste assimilation and flood attenuation (White et al. 2017). Risks to human health are accentuated in urban environments where populations are dense in precarious settlements and the likelihood of exposure to pollutants, floods, wetlands and diseases are high (Kjellstrom et al. 2007; Satterthwaite and Bartlett 2017). The cumulative presence of these health determinants generates favourable conditions for the emergence of diseases such as diarrhoea, cholera, acute respiratory infections, malaria, dengue fever, diabetes, arterial hypertension, cardiopathies, cancers. This is further exacerbated by poor performance of health systems that do not adequately respond to population care needs (WHO 2015).

In this perspective, the issues of urban environment and health crises refer to the continued exposure to infectious diseases risks and injuries associated with poor sanitation, unsafe drinking water, management of solid wastes, floods and wetlands, polluted air and toxic wastes in ecosystems, poverty and precarious habitats generating urban health problems and double burden of diseases (Hunter et al. 2010; Cissé et al. 2016). Despite these favourable conditions to the emergence of urban health risks, there is variability and heterogeneity of diseases epidemiology between settlements according to specific urban dynamics and contexts Salem (1994) and WHO (2017).

The complexity of urban health challenges in West African cities requires in depth analysis of spatial distribution of disease and the environmental conditions changing at different urban scales according to the type of settlements and neighbourhoods and the associated health risks (Salem 1998; Haggett 2000). Consequently, important socio-economic disparities have emerged in urban areas marked by profound health inequalities (Sy et al. 2010; Thiam et al. 2017a). Achieving the SDG 3 on ‘Health and well-being for all’ needs to deal with urban health issues using intersectoral approaches for making urban societies sustainable (Bartram and Cairncross 2010; UNICEF/WHO 2009).

The aim of this chapter is to analyse the spatial distribution of malaria and diarrhoeal diseases in West African cities through different research projects conducted between 2006 and 2016 in Mauritania, Senegal and Côte d’Ivoire. To highlight the geographical variation in urban health inequalities, a spatial epidemiology framework was applied in urban areas of Nouakchott and Kaédi (Mauritania), Pikine, Rufisque and Mbour (Senegal) and Korhogo (Côte d’Ivoire).

Spatial epidemiology is the study of spatial variation in disease risk and incidence. Several ecological processes can result in geographical disparities in health risk or disease incidence. For example, pathogen dispersal might be highly localized, vectors or reservoirs for pathogens might be spatially restricted, or susceptible hosts might be clumped (Ostfeld et al. 2005). In some instances, geography is implicit in spatial epidemiology (e.g. meta-population models of disease transmission), and sometimes explicit, (e.g. creation of risk maps and graphs for particular geographical areas) (Ostfeld et al. 2005).

Spatial Epidemiology: From Concept to Application in Urban Health

The epidemiological factors that control disease transmission operate over a range of spatial and temporal scales to produce spatially and temporally complex patterns of disease incidence (Nuckols et al. 2004). The importance of geographic information science is increasingly recognized in relation to spatial epidemiological research because it provides the fundamental geographic context to exploring spatial patterns in data (Rezaeian et al. 2007). In the context of epidemiology and public health, this provides a powerful aid to the analysis and understanding of the relationships between geography, the environment and human health. Spatial exploration methods are fast gaining importance and credibility in better understanding public and global health. These methods are particularly useful for attributing disease and risk to place, and targeting infectious disease interventions (Beale et al. 2008).

The British biomedical researcher, Jon Snow is largely hailed as the father of modern-day spatial epidemiology. His analyses of the geographic spread of cholera in London urban areas in 1843 led to a targeted intervention to stop the spread of the disease. In addition to Snow, the Russian parasitologist Pavlovsky's work on landscape epidemiology in the 1930s was 'discovered' by Western epidemiologists several decades later (Ostfeld et al. 2005). Following these first ideas, French and British geographers (Sorr 1933; Picheral 1983; Salem 1998; Haggatt 2000) developed several research project building in the concept of spatial epidemiology to help better understand the foundations and structures that explain the geographical distribution of disease, with particular emphasis on the role of natural factors. Progressively, the subsequent analyses of disease-environment interactions have highlighted the impact of human society's dynamics on the spatial distribution of health risks by integrating economic, political, social and cultural aspects (practices and behaviours) in order to replace the disease in its veritable spatial, social and environmental context (Sy 2006; Salem 1994).

The case studies presented in this chapter are built on spatial epidemiological approaches, using spatial analytical techniques and models to identify spatial anomalies (hotspots) in malaria and diarrhoeal diseases, according to the specificity of urban areas (Graham et al. 2004). Malaria transmission and interlinked factors were analysed in the cities of Pikine, Senegal (Cissé et al. 2016), Korhogo, Côte d'Ivoire and Kaédi, Mauritania, M'Bra (2017). These case studies mainly analysed implications of spatial epidemiology of urban malaria (vector-borne disease) by investigating urban environments that may provide potentials for malaria transmission, activity that exposes the vulnerable individuals to mosquito bites, and vegetation coupled with urban farming that provides ample aquatic habitats for mosquitoes (Machault et al. 2011; Dia et al. 2009; Faye et al 1995; Gadiaga 2011).

Diarrhoea is the second disease of interest, whose transmission and associated risk factors are demonstrated in the cities of Rufisque (Sy 2006) and Mbour (Thiam 2017) in Senegal and Nouakchott, Mauritania (Traoré 2016). In these regions of west Africa, several research projects on water, environmental sanitation and health have been conducted in urban areas. The case studies presented in this chapter mainly focused on spatial distribution of diarrhoeal diseases transmission around urban areas. Particular emphasis was placed on the essential components of risks factors known to be related to diarrhoea (Sambe et al. 2013; Sire et al. 2013).

Case 1: Spatial Epidemiology of Malaria in Pikine-Guédiawaye, Korhogo and Kaédi

Pikine is large city with a population exceeding 1 million inhabitants. The metropolitan area of Pikine-Guédiawaye has a concentration of some the most important of collections of water bodies, poverty, irregular habitats and informal agricultural practices. This area also has many flood prone areas and wetlands that provide an ecosystem conducive for vector breeding, exposing populations to malaria risk transmission. The methods used in this case study are based on the data analysis techniques from remote sensing, environmental and healthcare data and household surveys.

This study created maps that illustrate the geographical spread of the malaria in Pikine-Guédiawaye. The results showed a heterogeneous spatial distribution of confirmed malaria cases in the area with transmission rates significantly fluctuating among neighbourhoods that were in close proximity (Cissé 2017). Areas located near flooding areas and wetlands such as Ainouman Santa Yalla II (2.6%) and Demba Sow (2.4%) are among neighbourhoods with high prevalence, while incidence in Darou Rakhmane and Gazelle Ouest (1.5%) were moderate. A visual comparative analysis between the malarial case map and landcover map (Fig. 5.1) helps to reveal that malaria incidence decreases from flooding remote neighbourhoods to areas adjacent to vectors breeding sites. Incidence rates are lower in Grand Thiaroye (0.2%) and Gueule Tapée (0.4%) while many sectors have not recorded any cases of malaria such as Hamo I and HLM Paris (Fig. 5.2).

Korhogo is a medium-sized city with a population between 200,000 and 280,000 inhabitants whose urban context is characterized by poor access to basic social services, environmental sanitation problems, recurrent floods, water bodies (including dams and wells) and irrigated agriculture. The methods used in this case study were derived from environmental sanitation, climate, healthcare (morbidity) and household's surveys data.

The overall prevalence of *Plasmodium* was 13.30% (855/6430) while incidence of malaria was 18.30% (577/3153) in the rainy season compared with 8.48% (278/3277) in the dry season ($p < 0.001$). The incidence of malaria is therefore significantly higher in the rainy season than in the dry season. The results indicate that the risk of being infected with malaria is 1.4 times higher in women than in men (OR = 1.4, 95% CI = 1.07–1.79). In addition, the risk of infection is 1.3 times greater in households that do not use tap water for drinking (OR = 1.3, 95% CI = 1.00–1.82). It also appears that the lack of electricity in the home increases the risk of malaria by 1.6 (OR = 1.6, 95% CI = 1.17–2.12). The spatial distribution of disease at household scale showed that at least one person was infected by malaria during the rainy season and the dry season, in 2014 and 2015, respectively. The analysis of the results at the city level highlights the unequal distribution of malaria cases according to neighbourhoods and it appears that the central and eastern parts of the city are most affected by the disease. In contrast, the southern, western and extreme northern parts of the city (Teguéré, Résidentiel 1, Route Kapélé, Résidentiel 3 and Natiokobadara) were not affected as much in both seasons (Fig. 5.3).

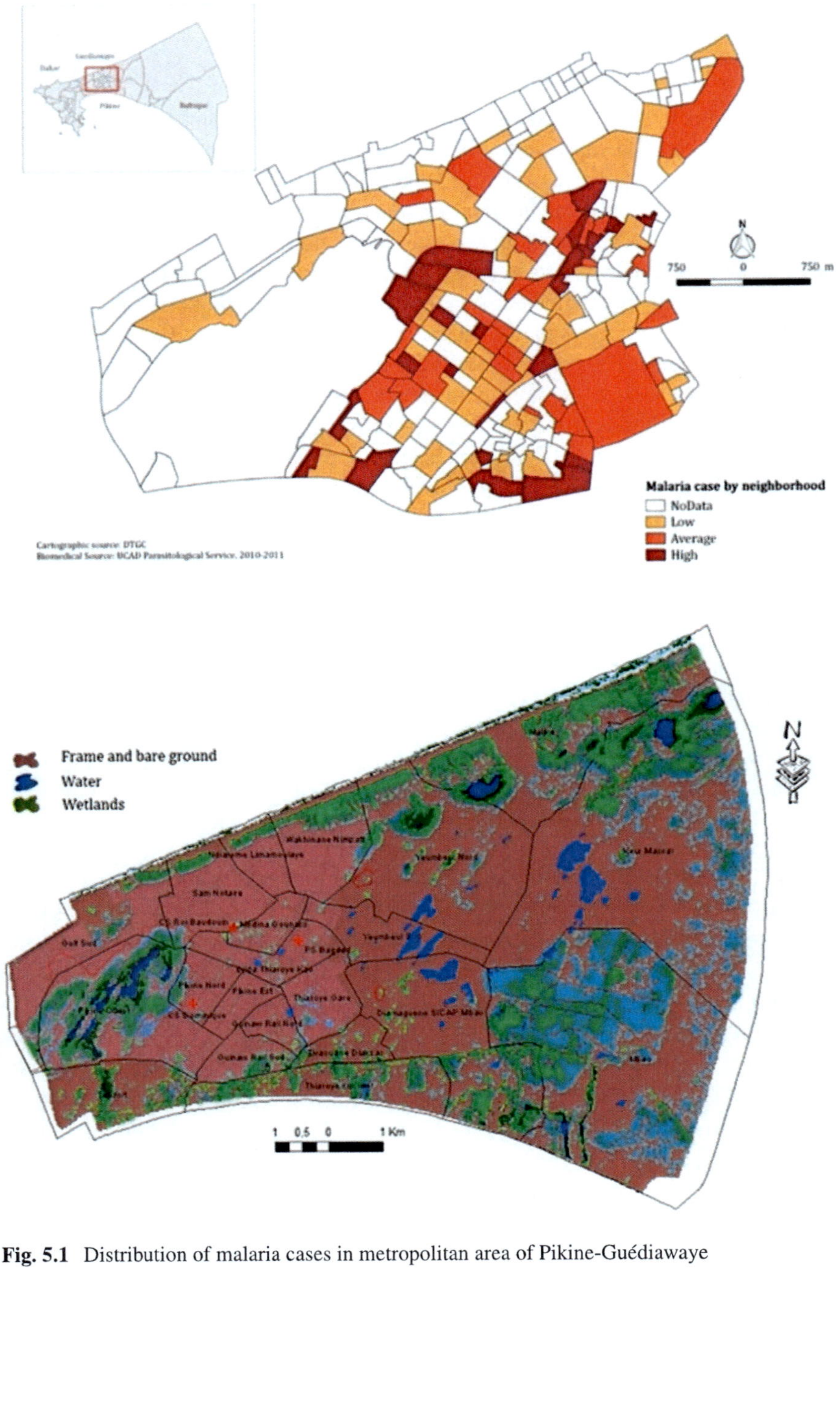

Fig. 5.1 Distribution of malaria cases in metropolitan area of Pikine-Guédiawaye

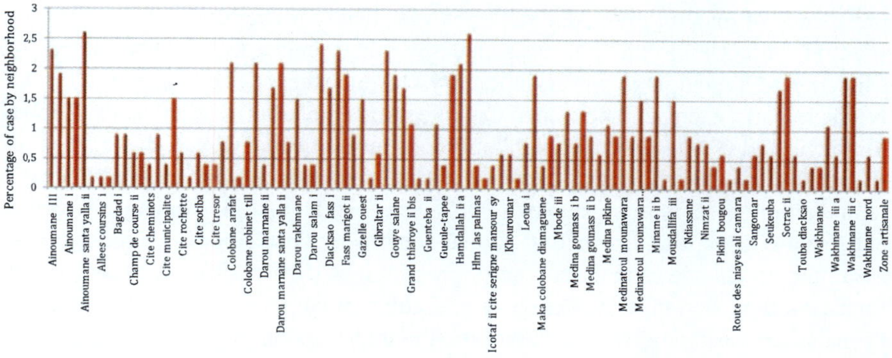

Fig. 5.2 Distribution of malaria incidence at neighbourhood scale in metropolitan area of Pikine-Guédiawaye

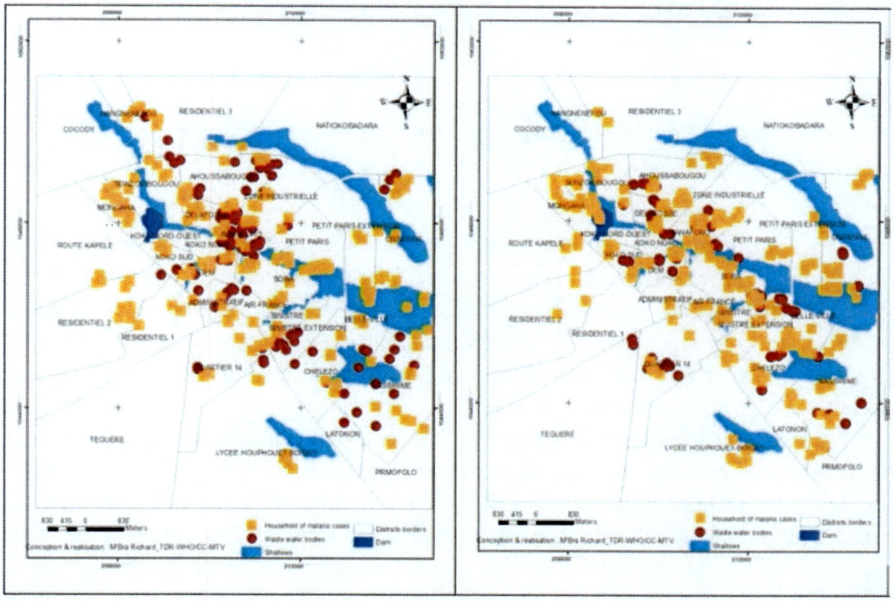

Fig. 5.3 Spatial distribution of malaria cases in Korhogo during the rainy and the dry seasons (parasitological survey, July 2014 and April 2015)

Kaédi is a small city with a population of 70,000–80,000 inhabitants whose urban context is marked by ecological ecosystem considered as favourable areas to mosquito larvae development. The methods used in this case study are based in the analysis of environmental sanitation, climate, healthcare (morbidity) and household's surveys data.

The overall prevalence of malaria in Kaédi was 0.3% (22/7702), composed of 0.07% (3/4002) in the rainy season and 0.51% (19/3700) in the dry season. The prevalence of malaria in the dry season is statistically higher than that in the rainy season ($p < 0.001$). In the rainy season, cases were found in the neighbourhoods of Modern and Tantadjil. However, in the dry season, except Tinzah and Inity (in the Northern part), each neighbourhood has at least one malaria case. In the city, Kebbe, Sinthiane and Gattaga are the neighbourhoods most affected (Fig. 5.4).

Because of the association between weather and climate patterns, particularly temperature, precipitation and extreme events, and malaria, climate change is expected to affect the distribution and seasonal activity of *Anopheles* mosquitoes, assuming no change in malaria control programmes (M'Bra et al. in press). In the focused cities, the range of average temperatures is sufficient for the development of malaria parasite and vector. Therefore, the transmission rate mostly depends on rainfall variation and the distribution of more specific environmental factors, potential breeding sites, such as potholes, small pits, cans or bins (Thwing et al. 2017). Another explanatory factor might be the extension of the cities into new neighbourhoods and the increase in urban agriculture contributing to increase the number of breeding sites and thus, malaria transmission (Lekweiry et al. 2015).

Case 2: Spatial Epidemiology of Diarrhoea in Rufisque, Nouakchott and Mbour

Rufisque is a medium-sized city with a population between 220,000 and 250,000 inhabitants whose urban context is characterized by poor access to basic social services, environmental sanitation problems, floods, water collections and irrigated open wastewater channels. The methods used in this case study are based in the analysis of water access and environmental sanitation, healthcare (morbidity) and longitudinal household's surveys data.

The spatial distribution of diarrhoea in the city of Rufisque showed an overall high incidence (6.5 episodes/child/year) in the study areas but large disparities are highlighted between the neighbourhoods observed according to the level of their environment sanitation. In 95.7% of households observed, diarrhoea cases were reported with an incidence ranging between 1 and 34.7 episodes per child/year. Although number of recorded episodes remains very important, there are large disparities between HLM district and other neighbourhoods (Castors, Diokoul Wague and Gouye Aldiana). Indeed, annual incidence rate per child is lower in the HLM district (sanitary: 3.4 episodes) than in Castors (little sanitary: 6.8 episodes), Diokoul Wague (unhealthy: 7.3 episodes) and Gouye Aldiana (very unhealthy: 8.4 episodes) (Figs. 5.5 and 5.6).

This high spatial variability is highlighted through univariate analysis which shows a significant association between the environmental profile of the district observed and the incidence of diarrhoea ($p = 0.0003$). Relative risk analysis shows that 37.3% of diarrhoeal risk is attributable to poor sanitation and hygiene ($p = 0.006$) and 24.5% of risk to precarious housing conditions. Household hygiene defective ($p = 0.01$). Similarly, socio-economic status contributes at 15.8% to diarrhoea risk

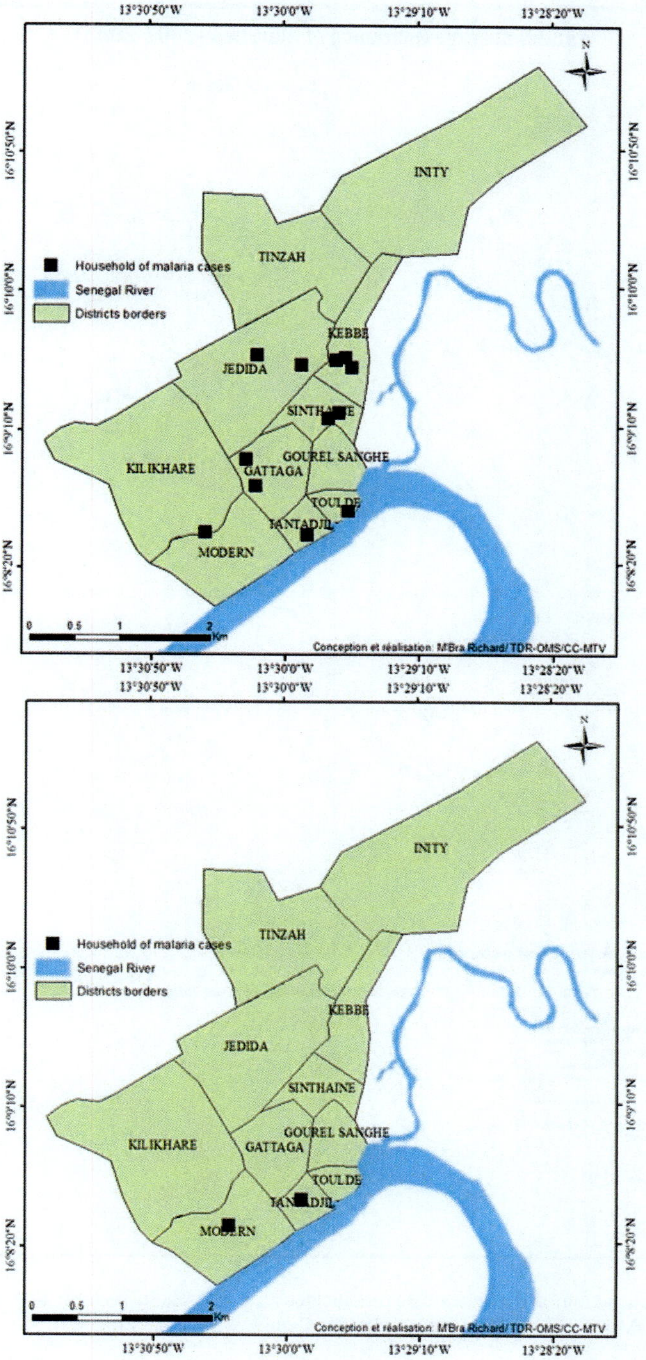

Fig. 5.4 Spatial distribution of malaria cases in Kaedi during the rainy season and in the dry season (parasitological survey, September 2014 and May 2015)

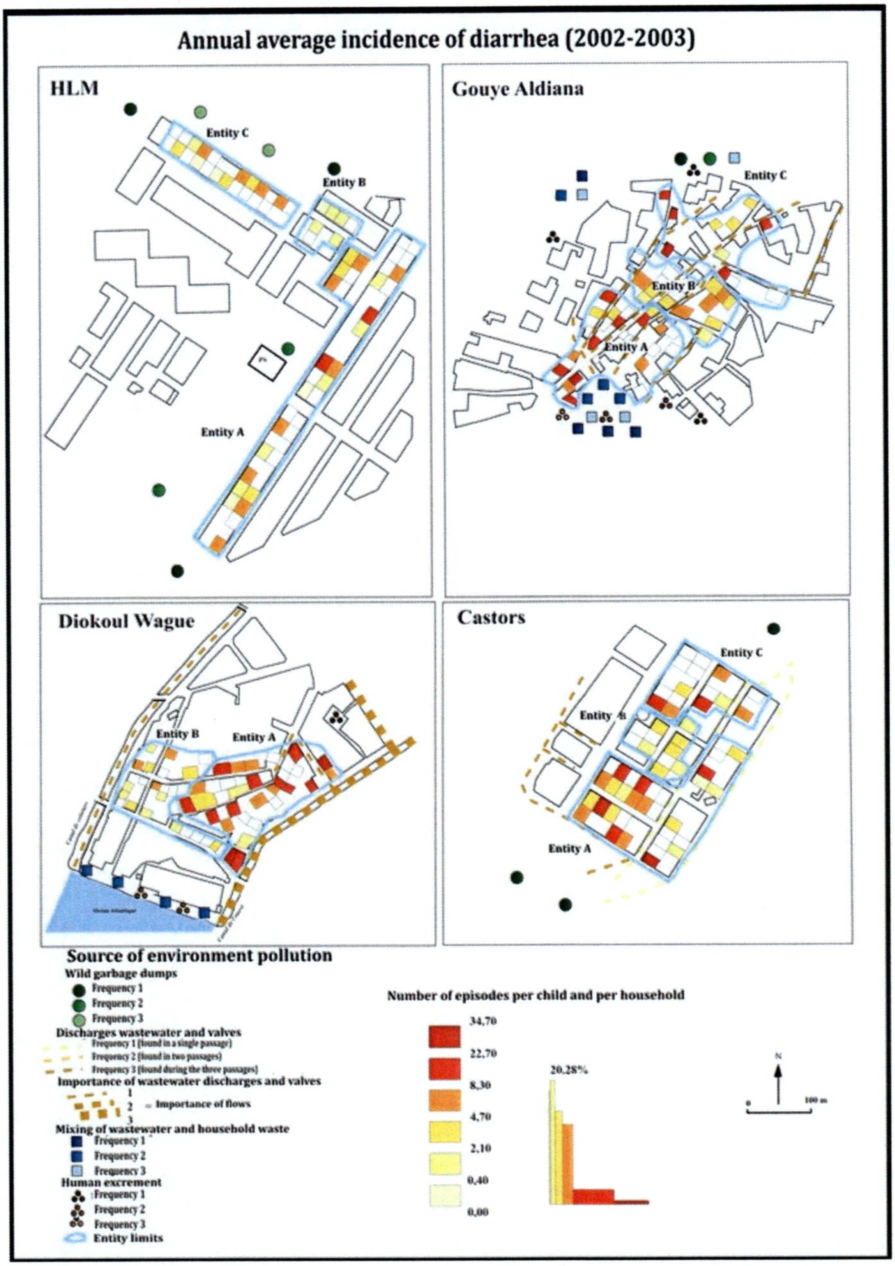

Fig. 5.5 Distribution of diarrhoea diseases incidence at household scale in Rufisque, Senegal. (Sources: ADM base map, 2001. Made with Phil Carto by Ibrahima SY, 2003)

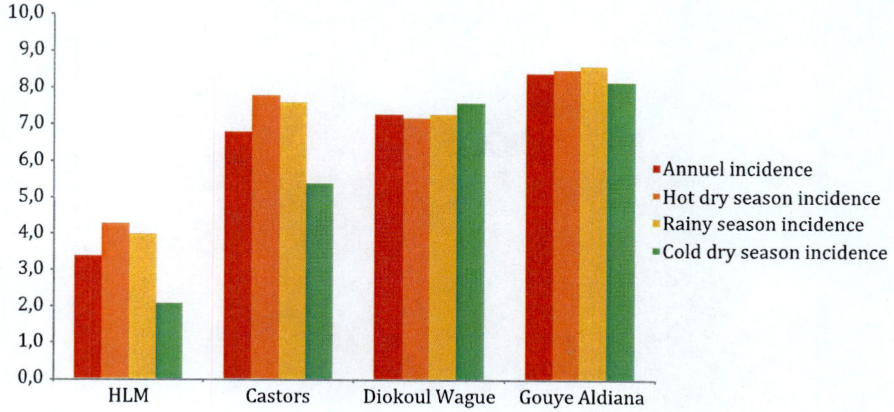

Fig. 5.6 Distribution of diarrhoea diseases incidence at neigbourhood scale in Rufisque, Senegal

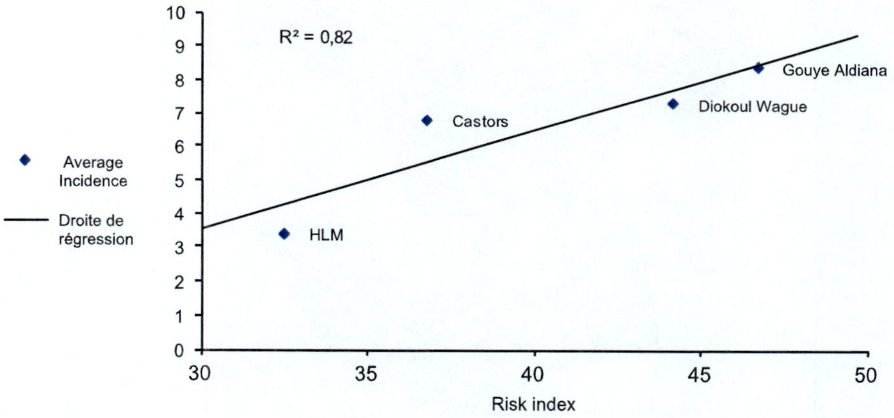

Fig. 5.7 Incidence of diarrhoea diseases at neigbourhood scale iaccording to standarded risk factors index in Rufisque, Senegal

($p = 0.04$) while 13.2% was attributable to poor household and individual's hygiene practices and behaviours ($p = 0.03$). The number of children per household was also a significant risk factor for which 9.2% of diarrhoeal risk was attributed ($p = 0.003$) (Fig. 5.7).

Nouakchott is a large city with a population exceeding 1 million inhabitants whose urban context is characterized by poor access to basic social services, environmental sanitation problems, recurrent floods, water collections and open air defecation. The methods used in this case study are based in the analysis of water access and environmental sanitation, healthcare (morbidity) and cross-sectional household's surveys data.

Diarrhoeal prevalence is 22.8% but morbidity is unequally distributed around urban areas ranging from 14.8% in Tevragh Zeina to 32.6% in Arafat (Fig. 5.8).

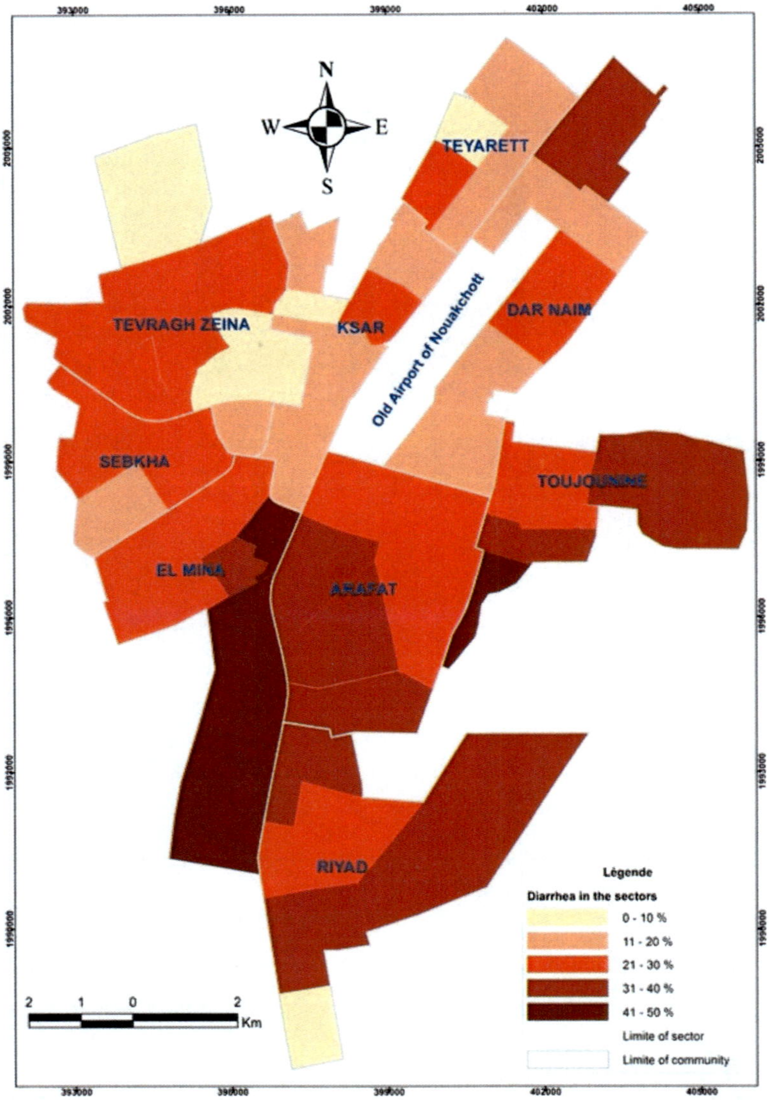

Fig. 5.8 Spatial distribution of diarrhoea prevalence at neighbourhood scale in Nouakchott. (Elaboration: Doulo Traoré (May 2018); Source map: MHUAT/DCIG, 2008 and CUN, 2008; Data source: NCCR-NS/RP14 survey (May 2012))

The municipalities with less prevalence average rounding 20% are Tevragh Zeina (14.8%), Ksar (15.2%), Teyarett (15.8%) and Dar Naim (19.8%) and the most affected are in importance order Arafat (32.6%), Toujounine (29.6%), Riyadh (28.6%) and El Mina (27.2%). Between the two groups which are differently exposed to diarrhoeal morbidity, there is the municipality of Sebkha with an preva-

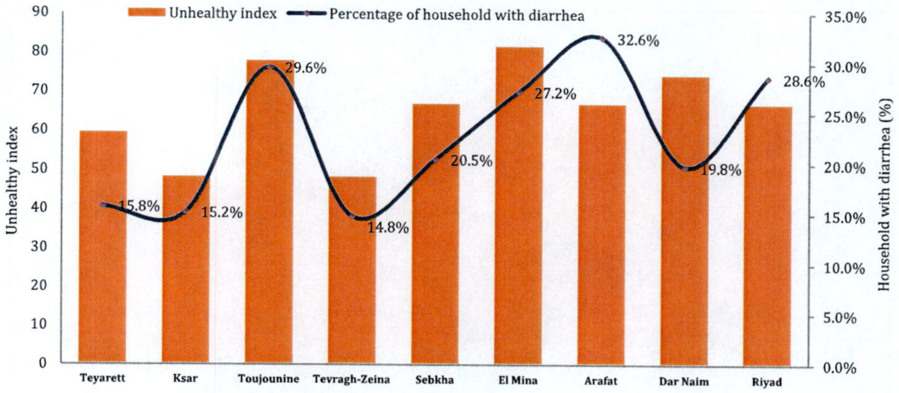

Fig. 5.9 Distribution of diarrhoea prevalence according to environmental sanitation at municipality scale in Nouakchott, Mauritania

lence average of 20.5%. The distribution of prevalence recorded according to age shows that children under 5 years are the most affected (16.5%) followed by the group aged 5–14 (7.2%) and adults (2.8%). If the analysis of diarrhoea epidemiological data at the inter-municipality scale shows an unequal distribution of the disease, there is a great spatial heterogeneity at the intra-municipality level between the sectors. Indeed, in El Mina, Arafat, Toujounine, Dar Naim and Ryadh municipalities are located sectors where the number of households affected by diarrhoea varies from 40% to 50% while those of Tevragh-Zeyna, Ksar and Teyarett have sectors with a number of affected households less than 10% in burden (Fig. 5.9).

The analysis of risk factors associated with the disease occurrence shows a significant correlation of diarrhoea prevalence with the environmental sanitation conditions ($p < 0.0001$), drinking water supply mode of households ($p = 0.018$), water storage at homes ($p = 0.003$), amount of water consumed ($p = 0.0002$), population density and household promiscuity ($p < 0.0001$), practice domestic animal breeding ($p = 0.0002$) and household health expenses ($p = 0.002$).

Mbour is a medium-sized city with a population between 270,000 and 280,000 inhabitants whose urban context is characterized by poor access to basic social services, environmental sanitation problems, floods, water collections and well waters. The methods used in this case study are based in the analysis of water access and environmental sanitation, healthcare (morbidity) and cross-sectional household's surveys data.

The diarrhoeal cases occurring within the 2 weeks preceding the interview were reported for one in four children, giving an overall prevalence of 26.1% ($n = 295$). Prevalence was slightly higher among girls than boys (27.6% and 24.4%, respectively), but this difference was not statistically significant ($p = 0.22$). Adjusted diarrhoea prevalence among children under 5 years did not show a significant difference between zones, with the highest rate observed in Urban Centre Area (26.9%) and the lowest rate in South peripheral area (17.1%). Without adjusting for other variables, the highest diarrhoea prevalence was observed in Peri-central area (44.8%) and the second highest in Urban Centre Area (36.3%) (Fig. 5.10).

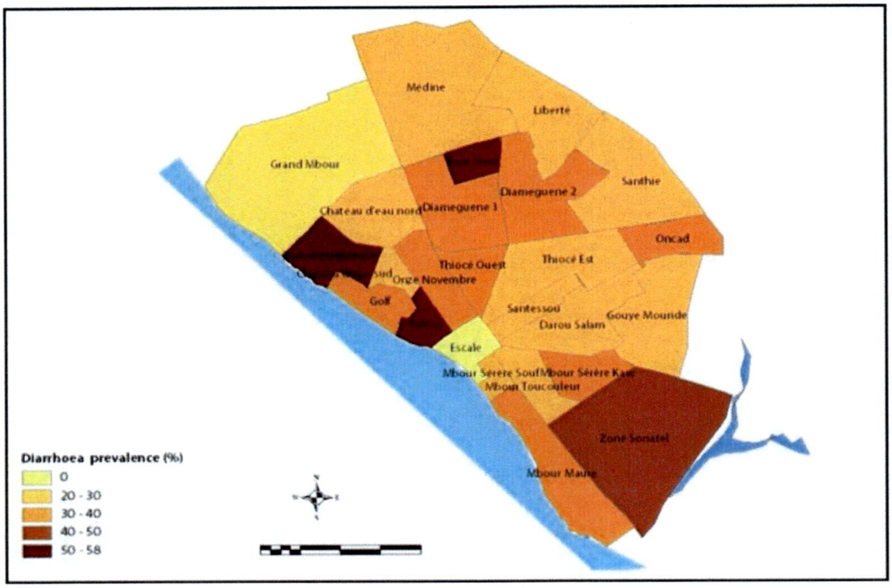

Fig. 5.10 Spatial distribution of diarrhoea prevalence at neighbourhood scale in Mbour, Senegal

The analysis stratified by age group showed a higher prevalence of diarrhoea in the oldest age group (24–59 months), while the lowest diarrhoea prevalence was observed in children under 1 year. Diarrhoea prevalence stratified by age group and zone is shown the highest prevalence among children aged 12–23 months in Peri-central area (Fig. 5.11).

Concerning household risk factors related to this disease, diarrhoea among children under 5 years was significantly associated with: (1) mother's unemployment (adjusted); (2) sharing the toilet with other households; (3) use of unconventional bag (open sack) for storing household solid waste; (4) households with more than one child under 5 years; (5) evacuation of household domestic wastewater in public street; and (6) no treatment of stored drinking water.

Main results show that diarrhoea risk in urban areas observed is strongly determined by the socio-ecological conditions (environmental sanitation, individual's practices and behaviours) and habitat environment (defective hygiene in households, population density, promiscuity, domestic breeding of animals). In the cities of Rufisque and Mbour, the seasonal effect of diarrhoea transmission is clearly observed at high spatial resolution, especially in neighbourhoods with a better environmental and sanitation condition (Thiam et al. 2017b; Sy et al. 2017). The lack of safe drinking water and poor sanitation and hygiene emerged as one of the main routes of diarrhoeal diseases transmission considered as major public health issue especially for children less than 5 years (Traoré et al. 2013; UNICEF-WHO 2015). Overall, the approach developed in spatial epidemiology has made it possible to identify factors, populations and areas at risk for diarrhoeal diseases and to highlight the challenges to be addressed by public health and hygiene policies (Fewtrell et al. 2005; Gatzweiler 2016).

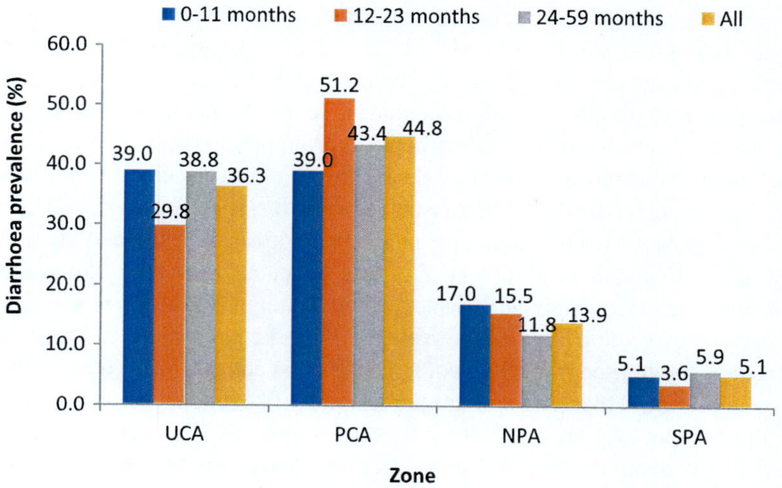

Fig. 5.11 Diarrhoea prevalence among children under 5-years-old by age and zone in Mbour, Senegal (period 2014). *UCA* urban central area, *PCA* peri-central area, *NPA* north peripheral area, *SPA* south peripheral area

Potential of Spatial Epidemiology for Public Health Improvements in Urban Areas

The analysis of unequal distribution of disease outcomes in relation with populations' environments makes it possible to question the emergence of health risk. Where health measures are normally presented at a national scale, multiplying case studies on the topic of the heterogeneity of disease transmission makes it possible to understand the real public health challenges in the urban areas. Findings of case studies in many African cities show that malaria and diarrhoea distribution can vary widely from one neighbourhood to another due to the diversity of living conditions in urban areas. This shows that data disaggregation across a geographical area such as cities does not necessarily make it possible to know the disparities existing between different sectors, neighbourhoods or cluster of houses composing them (Robinson 2000).

Spatial epidemiology enables one to determine the geographical and temporal distribution of the malaria disease and better understanding how environmental and social factors affect malaria transmission dynamics. This is an important first step in developing policies and programmes and planning interventions to target populations living in high-risk areas (Ostfeld et al. 2005). The diarrhoea analysis at urban scale does not allow to give complete picture of disease distribution in the cities and to highlight health inequalities between territories. Diarrhoeal epidemiology results at high spatial resolution compared to more aggregate levels has made it possible to highlight its spatial and social heterogeneity which will facilitate implementation of targeted interventions in most vulnerable communities (Traoré et al. 2013; Sy et al. 2017).

The need to better understand the contribution of other explanatory factors at both individual and neighbourhood level are vital to proposing future prevention strategies and interventions. Thus, the main findings from cases presented in this chapter illustrate the real and big potential of spatial epidemiology in addressing public health issues in African urban contexts. Mapping spatiotemporal dynamics of malaria and diarrhoea at different scales in the cities presented in this chapter enabled us to understand the factors that govern the spatial patterns and rate of spread of diseases. This highlights the challenges for urban health and the needs of interventions (Nuckols et al. 2004). The heterogeneous distribution of water and vector-borne diseases within an urban context is complex and requires location specific analysis to develop targeted interventions. This helps planners to assess spatial risk factors, and to ascertain what would be the most advantageous types of healthcare policies for the planning and implementation of healthcare services.

In this perspective, all inputs from the case studies in Mauritania, Senegal and Côte d'Ivoire demonstrated the importance of spatial epidemiologic research to improve urban health with more focused interventions in vulnerable communities in order to contribute to the achievement of the Sustainable Development Goals 3, 6, 11 and 13. The main findings show the unequal spatial distribution of malaria and diarrhoeal morbidity from the household to the municipality scale through clusters of houses, neighbourhood and sector or zone. The health inequalities that exist were exposed by the spatial epidemiology approach. This provides for intervention pathways to take into account the complexity and realities of urban health at different spatial scales.

Conclusion

The increased availability of spatial environmental, health and population data combined with improved statistical methods and spatial analysis techniques has fuelled an increase in spatial epidemiological studies, which assess the geographic distribution of potential health risks and their association with environmental risk factors. The ability to rapidly locate disease clusters, assess the spatial distribution of disease risk, and link environmental data and health outcomes provides a powerful tool for the evaluation of spatial relationships between disease and environmental conditions.

As demonstrated through results and findings from case studies in Mauritania, Senegal and Côte d'Ivoire urban contexts, spatial epidemiology helped to understand spatial heterogeneity of malaria and diarrhoea in urban areas with a particular emphasis on health risks associated with environmental hazards. Due to the nature of dominant health problems associated with rapid urbanization, public health issues can be addressed through the availability and the improvement of health determinants. Improving health outcomes will require intersectoral approach through the building of political alliance for urban health that involves stakeholders, urban planners, health officials and practitioners.

The scenario of future urban development in the context of high population growth, poverty increase and climate change is likely to lead to an amplification and aggravation of public health problems in cities. Therefore, it is important not only to increase the surveillance of environmental factors but also to promote health education to interrupt the transmission chain of infectious diseases. Furthermore, and interventions to strengthen health system performance for improving health and well-being of individuals in urban areas.

Acknowledgements The malaria cases presented in this chapter were funded by the International Development Research Centre (IDRC) and the 'Vulnerability and resilience to vector-borne disease in the context of climate change' supported by Research and Training in Tropical Diseases of the World Health Organization (TDR-WHO). The diarrhoea cases presented in this chapter were conducted as part of the framework of the Swiss National Centre of Competence in Research (NCCR) North-South and funded by the Swiss National Science Foundation and the Swiss Agency for Development and Cooperation (SDC).

References

Bartram, J., & Cairncross, S. (2010). Hygiene, sanitation, and water: Forgotten foundations of health. *PLoS Medicine, 7*(11), 9p.

Beale, L., Abellan, J. J., Hodgson, S., & Jarup, L. (2008). Methodologic issues and approaches to spatial epidemiology. *Environmental Health Perspectives, 116*(8), 7p.

Cissé, B. (2017). *Facteurs de risque du paludisme dans la banlieue de Dakar: Quels déterminants pour une meilleure prise de décision dans la lutte contre le paludisme?* PhD thesis in Health Geography, University Cheikh Anta Diop (UCAD), Dakar, Senegal.

Cissé, G., Traoré, D., Touray, S., Bâ, H., Keita, M., Sy, I., Koné, B., Utzinger, J., & Tanner, M. (2016). Vulnerabilities of water and sanitation at households and community levels in face of climate variability and change: Trends from historical climate time series in a west African medium-sized town. *International Journal of Global Environmental Issues, 15*(1/2), 81–99.

Dia, I., Ba, H., Mohamed, S. A. O., Diallo, D., Lo, B., & Diallo, M. (2009). Distribution, host preference and infection rates of malaria vectors in Mauritania. *Parasites & Vectors, 2*, 61.

Faye, O., Gaye, O., Fontenille, D., Hebrard, G., Konate, L., Ngayo, S. Y., et al. (1995). La sécheresse et la baisse du paludisme dans les Niayes du Sénégal. *Santé, 5*, 299–305.

Fewtrell, L., Kaufmann, R. B., Kay, D., Enanoria, W., Haller, L., et al. (2005). Water, sanitation and hygiene interventions to reduce diarrhoea in less developed countries: A systematic review and meta-analysis. *The Lancet Infectious Diseases, 5*, 42–52.

Gadiaga, L., Macault, V., Pages, F., Gaye, A., Jarjaval, F., Godefroy, L., Cissé, B., Lacaux, J. P., Sokhna, C., Trape, J. F., & Rogier, C. (2011). Conditions of malaria transmission in Dakar from 2007 to 2010. *Malaria Journal, 10*, 312. https://doi.org/10.1186/1475-2875-10-312. Retrieved from http://www.malariajournal.com/content/10/1/312/abstract.

Gatzweiler, F. W. (2016). *A systems approach to urban health and wellbeing to meet the sustainability challenges of urban change*. Brief for GSDR—2016 update. ICSU-UNU-IAMP Global, 4p.

Graham, A. J., Atkinson, P. M., & Danson, F. M. (2004). Spatial analysis for epidemiology. *Acta Tropica, 91*(2004), 219–225.

Haggett, P. (2000). *The geographical structure of epidemics*. Oxford: Oxford University Press, 149p.

Hunter, P. R., MacDonald, A. M., & Carter, R. C. (2010). Water supply and health. *PLoS Medicine, 7*(11), 9p.

IIED. (2016). *Informality and inclusive green growth: Evidence for the biggest private sector*. London: IIED.

Kjellstrom, T., Mercado, S., Sattherthwaite, D., McGranahan, G., Friel, S., & Havemann, K. (2007). Our cities, our health, our future: Acting on social determinants for health equity in urban settings. *Report to the WHO Commission on social determinants of health from the knowledge network on urban settings, Kobe, Japan*, 70p.

Lekweiry, K. M., Salem, M. S. O. A., Basco, L. K., Briolant, S., Hafid, J., & Boukhary, A. O. M. S. (2015). Malaria in Mauritania: Retrospective and prospective overview. *Malaria Journal, 14*, 100.

M'Bra, K. R. (2017). *Eau, assainissement et changement climatique: analyse intégrée des facteurs de risque au paludisme et aux bilharzioses à Korhogo (Nord de la Côte d'Ivoire) et à Kaédi (Sud de la Mauritanie)*. PhD thesis, University Felix Houphouêt Boigny of Cocodo, Abidjan, Côte d'Ivoire, 284p.

M'Bra, R. K., Kone, B., Soro, D. P., N'krumah, R. T. A. S., Soro, N., Ndione, J. A., Sy, I., Ceccato, P., Ebi, K. L., Utzinger, J., Schindler, C., & Cissé, G. (in press). Impact of climate variability on the transmission risk of malaria in northern Côte d'Ivoire. *PLoS One*.

Machault, V., Borchi, F., Vounatsou, P., Pagès, F., Vignolles, C., Lacaux, J.-P., & Rogier, C. (2011). The use of remotely sensed environmental data in the study of malaria. *Geospatial Health, 5*(2), 151–168.

Nuckols, J. R., Ward, M. H., & Jarup, L. (2004). Using geographic information systems for exposure assessment in environmental epidemiology studies. *Environmental Health Perspectives, 112*(9), 1007–1015.

Ostfeld, R. S., Glass, G. E., & Felicia, K. (2005). Spatial epidemiology: An emerging (or re-emerging) discipline. *Trends in Ecology and Evolution, 20*(6), 9p.

Picheral, H. (1983). Complexes et systèmes pathogènes: approche géographique», De l'épidémiologie à la géographie humaine. Travaux et documents de géographie tropicale, n° 48, p. 5–22.

Rezaeian, M., Graham, G., St Leger, S., & Appleby, L. (2007). Geographical epidemiology, spatial analysis and geographical information systems: A multidisciplinary glossary. *Journal of Epidemiology and Community Health, 61*, 98–102, 4p. https://doi.org/10.1136/jech.2005.043117.

Robinson, T. P. (2000). Spatial statistics and geographical information systems in epidemiology and public health. *Advances in Parasitology, 47*, 81–128.

Salem, G. (1998). *La santé dans la ville. Géographie d'un espace dense: Pikine (Sénégal)*. Paris: Editions Karthala-ORSTOM; 360p.

Salem, G., et al. (1994). Espace urbain et risque anophélien à Pikine (Sénégal). *Cahiers Sante, 4*, 347–357.

Sambe, B., Espie, E., Faye, M. E., Timbine, L. G., Sembene, M., & Gassama-Sow, A. (2013). Community-acquired diarrhea among children and adults in urban settings in Senegal: Clinical, epidemiological and microbiological aspects. *BMC Infectious Diseases, 13*, 580.

Satterthwaite, D., & Bartlett, S. (2017). Editorial: The full spectrum of risk in urban centres: Changing perceptions, changing priorities. *Environment and Urbanization, 29*(1), 3–14.

Sire, J. M., Garin, B., Chartier, L., Fall, N. K., Tall, A., Seck, A., Weill, F. X., Breurec, S., & Vray, M. (2013). Community-acquired infectious diarrhoea in children under 5 years of age in Dakar, Senegal. *Paediatrics and International Child Health, 33*(3), 139–144.

Sorr, M. (1933). Complexes pathogènes et géographie médicale, *Annales de Géographie*, 1933, vol. 42, n°235, p. 1–18.

Sy, I. (2006). *La Gestion de la Salubrité à Rufisque (Sénégal): Enjeux Sanitaires et Pratiques Urbaines*. PhD thesis in Health Geography, University Louis Pasteur of Strasbourb, France, 563p.

Sy, I., Handschumacher, P., Wyss, K., Piermay, J. L., Cissé, G., & Tanner, M. (2010). Variabilité des maladies diarrhéiques chez les enfants de moins de 5 ans en milieu urbain: l'exemple de Rufisque au Sénégal. *Médecine Tropicale, 70*, 57–62.

Sy, I., Traore, D., Niang Diene, A., Koné, B., Epprecht, M., Ndione, J. A., Lô, B., Bonfoh, B., Toure, A., Cisse, G., & Tanner, M. (2017). Water, sanitation and diarrheal diseases in Nouakchott

(Mauritania): Analyzing the determinant of urban health inequalities. *Journal of Infectious Diseases and Preventive Medicine, 6*, 172. https://doi.org/10.4172/2329-8731.1000172.

Szombathely, M., Albrecht, M., Antanaskovic, D., Augustin, J., et al. (2017). A conceptual modeling approach to health-related urbanwell-being. *Urban Science, 1*, 17. https://doi.org/10.3390/urbansci1020017. Retrieved from www.mdpi.com/journal/urbansci.

Thiam, S. (2017). *Spatial analysis of diarrhoea and environmental risk factors in an urban context in Senegal, West Africa*. PhD thesis in Epidemiology and Public Health, Swiss Tropical and Public Health Institute (Swiss TPH), University of Basel, Switzerland.

Thiam, S., Diène, A. N., Fuhrimann, S., Winkler, M., Sy, I., Ndione, J. A., Schindler, C., Vounatsou, P., Utzinger, J., Faye, O., & Cissé, G. (2017a). Prevalence of diarrhoea and risk factors among children under five years old in Mbour, Senegal: A cross-sectional study. *Infectious Diseases of Poverty, 6*, 109. https://doi.org/10.1186/s40249-017-0323-1.

Thiam, S., Fuhrimann, S., Niang-Diène, A., Sy, I., Faye, O., Utzinger, J., & Cissé, G. (2017b). Urbanisation and its effect on risk factors associated with childhood diarrhoea in Mbour, Senegal: A visualization. *Geospatial Health, 12*, 632. Retrieved from https://www.youtube.com/watch?v=DRQJgu2WiXg&t=5s.

Thwing, J., Eckert, E., Dione, D. A., Tine, R., Faye, A., Yé, Y., Ndiop, M., Cisse, M., Ndione, J. A., Diouf, M. B., & Ba, M. (2017). Declines in malaria burden and all-cause child mortality following increases in control interventions in Senegal, 2005–2010. *American Journal of Tropical Medicine and Hygiene, 97*(3_Suppl), 89–98. https://doi.org/10.4269/ajtmh.16-0953.

Traoré, D.(2016). *Eau, environnement et santé dans un écosystème côtier urbain: approche géographique à Nouakchott (Mauritanie)*. PhD thesis in Health Geography, University Cheikh Anta Diop (UCAD), Dakar, Senegal, 263p.

Traoré, D., Sy, I., Utzinger, J., Kengne, I. M., Epprecht, M., Lô, B., Faye, O., Bonfoh, B., Cissé, G., & Tanner, M. (2013). Water quality and health in Sahelian semi-arid urban context: Geographical approach in Nouakchott (Mauritania). *Geospatial Health, 8*(1), 53–63.

UNICEF/WHO: Diarrhoea. (2009). *Why children are still dying and what can be done*. The United Nations Children's Fund (UNICEF)/World Health Organization (WHO).

UNICEF-WHO. (2015). *25 Years progress on sanitation and drinking water*. 2015 Update and MDG Assessment, Joint Monitoring Programme (JMP), 90p.

White, R., Turpie, J., & Letley, G. (2017). *Greening Africa's cities: Enhancing the relationship between urbanisation, environmental assets and ecosystem services*. Washington, DC: World Bank.

WHO. (2015). *World malaria report 2015*. Geneva: World Health Organization. Retrieved from http://www.who.int/mediacentre/factsheets/fs094/en/.

WHO. (2017). *World health statistics 2017: Monitoring health for the SDGs, sustainable development goals*. Geneva: World Health Organization.

WHO-UN Habitat. (2010). *Hidden cities: unmasking and overcoming health inequalities in urban settings*. Geneva: WHO-UN-Habitat Report, 145p.

Chapter 6
Methods of Measuring Spatial Accessibility to Health Care in Uganda

Paul Ouma, Peter M. Macharia, Emelda Okiro, and Victor Alegana

Abstract Ensuring everyone has access to health care regardless of demographic, geographic and social economic status is a key component of universal health coverage. In sub-Saharan Africa, where populations are often sparsely distributed and services scarcely available, reducing distances or travel time to facilities is key in ensuring access to health care. This chapter traces the key concepts in measuring spatial accessibility by reviewing six methods—Provider-to-population ratio, Euclidean distance, gravity models, kernel density, network analysis and cost distance analysis—that can be used to model spatial accessibility. The advantages and disadvantages of using each of these models are also laid out, with the aim of choosing a model that can be used to capture spatial access. Using an example from Uganda, a cost distance analysis is used to model travel time to the nearest primary health care facility. The model adjusts for differences in land use, weather patterns and elevation while also excluding barriers such as water bodies and protected areas in the analysis. Results show that the proportion of population within 1-h travel times for the 13 regions in the country varies from 64.6% to 96.7% in the dry period and from 61.1% to 96.3% in the wet period. The model proposed can thus be used to highlight disparities in spatial accessibility, but as we demonstrate, care needs to be taken in accurate assembly of data and interpreting results in the context of the limitations.

Keywords Spatial access · Cost distance analysis · Primary health facilities · Travel times · Disparities in access

P. Ouma (✉) · P. M. Macharia · E. Okiro · V. Alegana
Population Health Unit, KEMRI-Wellcome Trust Research Programme, Nairobi, Kenya
e-mail: Pouma@kemri-wellcome.org

P. T. Makanga (ed.), *Practicing Health Geography*, Global Perspectives on Health Geography, https://doi.org/10.1007/978-3-030-63471-1_6

Introduction

Access to health is a key component of universal health coverage, which aims to ensure that all individuals are able to obtain quality health services regardless of their demographic and socio-economic status (Evans et al. 2013). Access is defined as the opportunity or ease with which consumers or communities are able to use appropriate services in proportion to their needs (Ensor and Cooper 2004). Poor access to health care services has been identified as a challenge in many countries in sub-Saharan Africa (SSA). Consequently, vulnerable populations, particularly children and mothers die from illness and conditions whose interventions are available at health facilities (Feikin et al. 2009; Rutherford et al. 2009, 2010; Schoeps et al. 2011; Okwaraji et al. 2012). Therefore, analysis of variation in geographical access to care is important, not only as an indicator of the strength of a health system but also to identify vulnerable populations at greater risk of preventable diseases.

Health care access is multi-dimensional and entails availability, acceptability, accommodation, affordability and accessibility. Availability concerns itself with resources available in delivering an intervention such as characteristics of health facilities (density, distribution and decentralization) and those affecting utilization such as duration and flexibility of hours of operation. Acceptability refers to the patient's interaction with health care systems in terms of choice based on such factors as gender, culture and the perception of the provider towards the patient in terms of age, social class and ethnicity. Accommodation is the arrangement and organization of health services in order to meet population demand while affordability is the population's ability to meet financial obligations related to medical services (Aday and Andersen 1974; Penchansky and Thomas 1981; Alun and David 1984; Higgs 2004; Haas et al. 2004; Levesque et al. 2013; Gautam et al. 2014). The final dimension is the accessibility commonly referred to as spatial or geographic accessibility to health care. These are summarized in Fig. 6.1.

In SSA, populations are often sparsely distributed with few health facilities. Transport infrastructure is often poor, hence the influence of geography may overshadow the other aspects of access. The focus of this chapter is on geographic accessibility. Traditionally, unavailability of tools/software and datasets that can account for geographic factors such as land use, elevation and road networks that affect transport have been significant bottlenecks in accurately defining geographic access to health care. However, in the recent past there has been increased development of geographic analysis tools able to offer sophisticated analysis allowing for the development of different accessibility models (Neutens 2015). Additionally, spatially disaggregated data are becoming increasingly available at a higher spatial and temporal granularity.

The chapter discusses different approaches of modelling spatial access to health care including their advantages, limitations and data needs. The methods include provider-to-population ratio, distance and travel time metrics based on Euclidean, road network and cost distance algorithms. Based on the discussion, one method is used to demonstrate the implementation of geographic access modelling in Uganda as a case study.

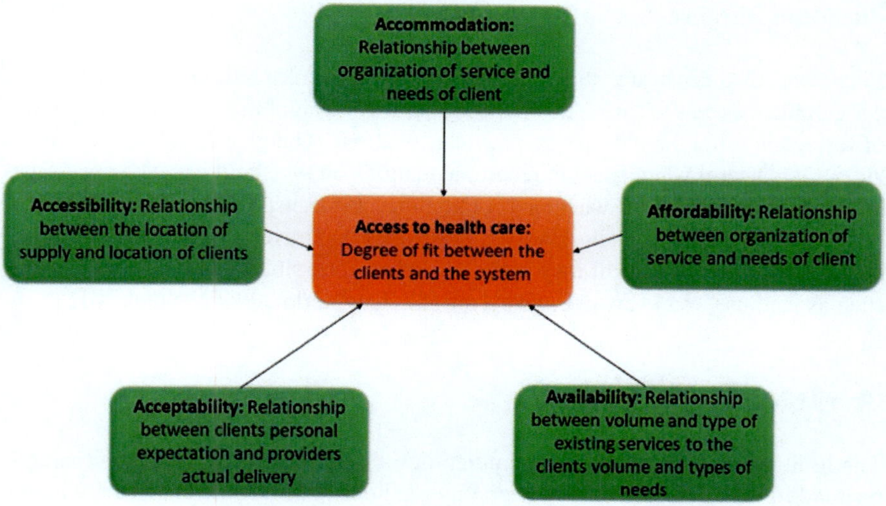

Fig. 6.1 The multi-dimensional concept of health care access

Geographic Accessibility

Geographic access to health care refers to the difficulty or ease in moving from a place where a need for health services is triggered to where the health service provider is located. It addresses the complex interactions between population distribution, location of services and how people move to the health services. The three common ways of measuring geographic access are the provider versus people (need), distance or by travel time, and the methods used to estimate these are provided in the next section.

Measuring Geographic Accessibility

Provider-to-Population Ratio

This method involves calculating the provider-to-population ratio (PPR) (Neutens 2015) based on the number of health facilities, doctors, number of beds, etc., on a predefined administrative area relative to the populations in these areas. PPR is useful in highlighting differences between administrative boundaries and identification of gaps in service provision (WHO 2010). The technique is simple and does not necessarily require Geographic Information Systems (GIS) skills. However, it does not account for travel impedances (elevation, transport availability and distance) encountered when accessing health facilities (Neutens 2015).

Euclidean Distance

Euclidean distances are the simplest distance-based method for quantifying geographical access to care. The method assumes a straight line of travel from points of residence to the health service provider locations (Guagliardo 2004; Noor et al. 2009). It is useful when specific recommendations on threshold distance exist especially in the in rural areas where access to motorized transport is limited and lack of health facilities are minimal. However, it assumes that travel occurs in a straight line and ignores the influence of transport services on accessibility and barriers of travel such as land use, road network and elevation (Guagliardo 2004; Neutens 2015).

Gravity Models

The limitations of the facility to population ratio and Euclidean distance methods resulted in the development of the gravity methods (Luo 2004). The gravity model is a combination of availability and accessibility across defined spatial units. It controls for "capacity" of a facility, competition between facilities and ability to estimate gravity values using numerous methods (Neutens 2015). Capacity here refers to the number of patients a facility can handle, which can be a function of the staffing and equipment available. The incremental developments in the model have seen it evolve from simply using the supply and demand data to the inclusion of distance decay effects, multiple transport models and variable incorporation of catchment areas in the modified two-step floating catchment area methods (McGrail and Humphreys 2009; Wan et al. 2012; Hu et al. 2013; Mao and Nekorchuk 2013; Vora et al. 2015). As such, the models have evolved from simply defining two-step floating catchment area (2SFCA) to the more sophisticated modified two-step floating catchment area (M2SFCA) method (Delamater 2013; Ni et al. 2015).

However, limitations still exist in the model, mainly its static nature and inability to allow for time varying relationships. Secondly, demand is normally defined at specific spatial units and the model would be affected by the Modifiable Areal Unit Problem (MAUP), a source of spatial bias which results from the aggregation of data. Its accuracy is therefore dependent on the ability to define population at fine geographic units and availability of data on service provider capacity. It is therefore not always suitable for use in resource limited settings where populations are normally defined at large spatial units.

Kernel Density Method

The kernel density model is a variant of the gravity model, which operates by distributing a discrete point value in a surface that is continuous (Schuurman et al. 2010). Kernel density is a non-parametric way of representing the distribution of a variable and allows estimation of a probability density function randomly. With regards to health service provision, a kernel density around a health service provider represents a 'sphere of influence' whose radius is the bandwidth of the kernel density. This method is limited in several ways; it uses straight line distances ignoring the road networks, which affect the ability to access a health facility. Secondly, its arbitrariness in choosing the kernel density in most cases leads to service densities that spill over from the study area (Guagliardo 2004). Thirdly, when modelling the population distribution, the method assumes a smooth distribution from a centroid with density decreasing as distance from the centroid increases, an assumption which is not realistic (Schuurman et al. 2010).

Network Analysis

Network analysis entails the use of the actual transport/travel routes to compute either travel time or distance to the nearest service provider (Noor et al. 2006; Owen et al. 2010; Masoodi and Rahimzadeh 2015). It is superior to the Euclidean method in this regard (Tansley et al. 2015). Although it's a more realistic method, its usability in rural areas may be affected by the fact that transport does not always follow the road network (Nesbitt et al. 2014). Accurate data on transportation routes and populated locations is also difficult to obtain. The algorithm assumes travel can only occur along the roads and it is a more computationally intensive method which relies on the ability to define population locations (nodes), accurate transport infrastructure and the routes likely to be used.

Cost Distance Analysis

Cost distance techniques provide more intuitive methods of defining accessibility for policy makers (Guagliardo 2004; Noor et al. 2006; Wang 2012; Nesbitt et al. 2014) because travel times are more realistic representations of access as people relate more to the time taken to get to a health facility than to distances. The availability of datasets that can be used to define travel times in recent time, makes it a more attractive choice of defining accessibility. It involves the development of a 'cost surface' that defines travel speeds within different land covers, roads and elevation. This surface is then used in combination with the location of health facilities in a 'cost distance' analysis to come up with a surface showing the least time needed to get to each health facility for every populated location (Ray and Ebener 2008; Huerta and Källestål 2012). Limitations of this method include the assumption that individuals use the nearest facility and its inability to account for competition (Neutens 2015). In addition, its accuracy is dependent on the spatial resolution used. However, it is very useful in SSA, because the influence of factors such as competition and choice are often overridden by distance (Neutens 2015).

Its development has gained significant traction with the development of the WHO AccessMod module for measuring physical accessibility (Ray and Ebener 2008) and other open source modules used in computational platforms like R. AccessMod for example is a standalone module that is easy to use, requiring basic GIS knowledge and is freely available (www.accessmod.org). This method was used in this study to assess variation in geographical access to care in Uganda.

Spatial Access to Primary Health Care: A Case Study of Uganda

Spatial Databases

Fundamental to estimating spatial access and marginalized population are spatially defined databases. We discuss the key datasets required.

Health Facility List

To estimate spatial access to health care, an authoritative, complete facility list is required. Key variables needed are facility name, unique identifier, location, facility type, ownership and operational status. Only a handful of countries in SSA have updated facility list (Noor et al. 2004, 2009; Rose-Wood et al. 2014; MEASURE Evaluation 2018) and health facility lists in SSA remain fragmented (WHO 2012; USAID, WHO 2018). The first inventory of pan African public health facility list was assembled using a disparate list of sources from national and international organizations (Maina et al. 2019). This exercise involved triangulating between different sources to check and remove duplicates, geocode, confirm spatial locations and administrative boundaries while validating facility numbers with those reported in health sector strategic reports. The final list is publicly available (Maina et al. 2019) and provides the most comprehensive publicly available resource of health facilities in SSA and Uganda as shown in Fig. 6.2.

Accessibility Covariates

To model spatial accessibility in a cost distance algorithm, additional covariates that define physical barriers of access are needed. Land cover data at 20 m spatial resolution was obtained from the RCMRD data portal (RCMRD 2017), that was produced by classifying remotely sensed data from the Sentinel-2 sensor. Publicly available road network data was obtained from the OpenStreetMaps and Google Map Maker projects. A digital elevation model raster surface (DEM) was downloaded from NASA's Shuttle Radar Topography Mission at the USGS Land Processes Distributed Active Archive Center (LP DAAC) website at 30 m spatial resolution. Auxiliary data in the form of water bodies and rivers were assembled. The water bodies are considered non-traversable expect where road bridges are constructed and/or use boats and canals were applicable. These were downloaded from the Global Lakes and Wetlands Database (Lehner and Döll 2004).

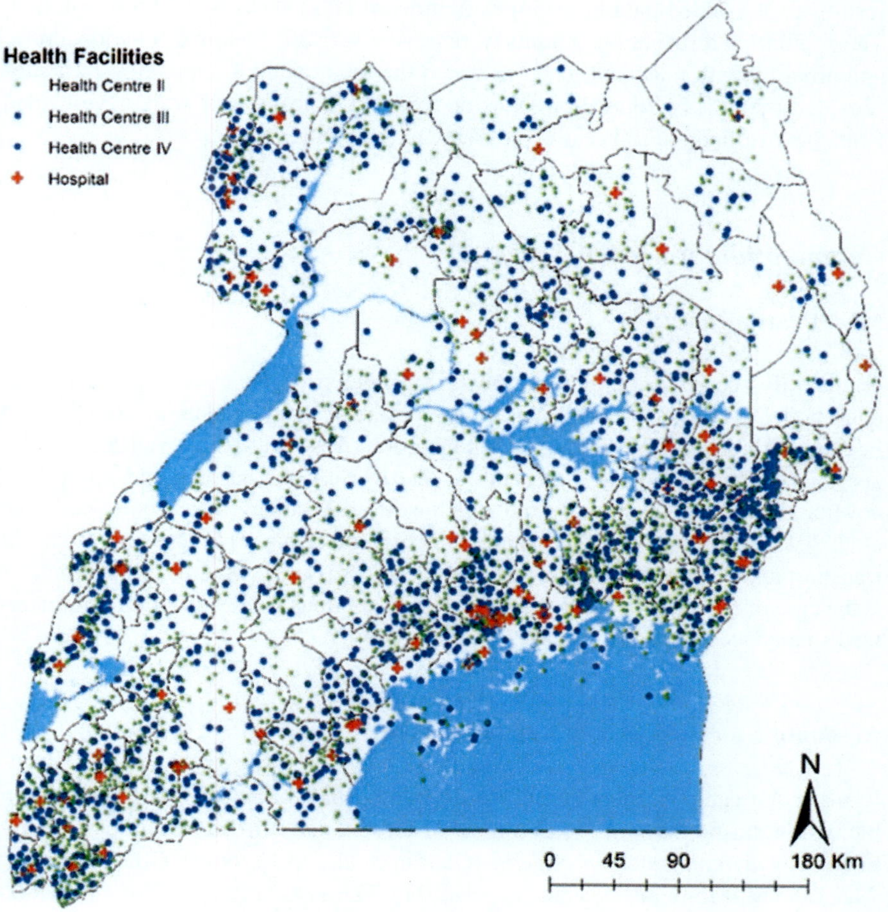

Fig. 6.2 3771 public health facilities including Health Centre II (2217), Health Centre III (1242), Health Centre IV (191) and Hospitals (121)

Population Distribution

Spatial modelling techniques for the reallocation of populations within census units have been developed in an attempt to overcome the difficulties caused by input census data of varying, and often low, spatial resolutions (Linard et al. 2012). A dasymetric modelling technique (Mennis 2009) was used to redistribute population counts within the 6255 spatially defined Parishes (sub national units representative of Administrative level 4 unit) used during the 2002 national census and land cover datasets derived from satellite imagery. Covariates used were land use, night time lights, water bodies, protected areas and elevation. A different population weight was assigned to each land cover class in order to shift populations away from unlikely populated areas and concentrate populations in built-up areas. The net result was a gridded dataset of population distribution at 0.1 × 0.1 km resolution. The population distribution datasets were projected to 2015 using UN national rural and urban growth rates and made to match the total national population estimates provided by the UN Population Division for 2015. The datasets were downloaded from the Worldpop portal accessible freely to the public.

Computation of Spatial Access

Model Parametrization

To quantify spatial accessibility to health facilities in Uganda we adopted a raster data model (Delamater et al. 2012) based on the geography, data availability and health-seeking behaviour of the study population (Nesbitt et al. 2014). Raster-based approach is more likely to represent the real world since barriers, elevation, land cover and roads can be incorporated with the different modes of transport (Rodrigue et al. 2013). A daunting task is usually to assign travel speeds to different modes of transport and different road types and land cover classes. With no empirical data as is the case of most SSA countries, we reviewed previous studies that had parameterized spatial access models in SSA (Table 6.1).

Accounting for the Effect of Rainfall Seasonality

In many African countries, during the dry season most unpaved roads are accessible, while during the rainy season most of these unpaved roads are impassable as previously demonstrated in Nigeria (Okafor et al. 2009), Niger (Blanford et al. 2012) and Mozambique (Makanga et al. 2017). This is usually caused by substantial rainfall and flooding in the wet season. Similar analyses have been conducted in Mozambique (Makanga et al. 2017), where daily flood extent raster layers and precipitation data were used to mark inaccessible areas during wet seasons. Reduced speeds were recorded where the daily precipitation was above 1 mm. This allowed the effect of seasonal variation in spatial access to be viewed as continually varying as compared to a binomial coding of wet and dry season (Makanga et al. 2017). We demonstrate the effect of the wet and dry season, by running two models with different parametrization speeds based on the season. The speeds assigned to different road categories during the wet season were 80% of those assigned during the dry season (Makanga et al. 2017) as shown in Table 6.1.

Table 6.1 Travel speeds across different land cover sources

Category	Class	Mode of travel	Speed in km/h (dry season)	Speed in km/h (wet season)
Land use land cover	Tree cover areas	Walking	2	1.6
	Shrub cover areas	Walking	5	4
	Grassland	Walking	4	3.2
	Cropland	Walking	5	4
	Regularly flooded	Walking	0.01	0.01
	Sparse vegetation	Walking	4	3.2
	Bare areas	Walking	2	1.6
	Built-up areas	Walking	5	4
	Open water	Walking	0.01	
Roads	Major arterial	Motorized	50	40
	Minor arterial	Motorized	30	24
	Primary highway	Motorized	50	40
	Secondary road	Motorized	30	24
	Terminal	Bicycling	10	8

Source: Noor et al. (2006), Alegana et al. (2012), Blanford et al. (2012), Dixit et al. (2016), Macharia et al. (2017a, b), Makanga et al. (2017), Ouma et al. (2017, 2018)

Computation of Travel Time

A cost distance algorithm that modelled a composite of walking, bicycling and motorized travel time to the nearest public health service provider was used. A travel impedance surface was generated by assigning travel speeds in Table 6.1 for wet and dry seasons independently. We used the impedance surface and location of public health facilities to estimate time in minutes needed to travel to the nearest public health facility in AccessMod (version 5). The DEM was used to apply a correction for slope while walking using Tobler's law (Tobler 1993). Tobler's formulation decreases the up-slope walking speed as the slope increases, while slightly increasing the speed for a slightly negative slope when walking down-slope. The bicycling power correction was applied for the bicycling mode of travel (Zorn 2008; Austin 2012). The lakes and rivers were treated as a barrier and considered impassable. The time needed to get to a facility was determined by cumulatively adding the time needed to cross contiguous pixels in the so-called least cost path from any location in Uganda to the nearest facility. We used 100 m spatial grids to capture finer heterogeneity in travel times since the algorithm converts roads to a raster surface thereby affecting the accuracy of the model.

Mapped Travel Time and Population Coverage

The result of computing travel time to the nearest public health facility during the dry and wet seasons are shown in Fig. 6.3. The southern part of the country relative to North had low travel time to the nearest facility. The gridded surfaces from the dry and wet seasons models showed a similar pattern of variability across the country.

The population within defined travel time thresholds were estimated by overlaying the population grids together with the travel surfaces for both wet and dry seasons by regions shown in Table 6.2. Over 84% and 96% of the country's 2015 population live within 1 and 2 h of the nearest public health facility during the dry season respectively. There exists great variability at the regional level ranging between 74.1% and 91.9% in respect to the population within an hour of the nearest facility. Our models suggested that those who are not able to access health care within 1 and 2 h possibly due to the effect of the roads being affected by the rainfall was 2% and 0.6% respectively at the national level and variable across the regions (Table 6.2).

Conclusion

This chapter discussed accessibility methods and used a country case study in Uganda to illustrate some of the challenges involved in modelling spatial accessibility, using a procedure that relies on open source tools and datasets. The usefulness of this model is its ability to capture complexities in travel across different landscapes including the influence of weather, all which often act as barriers to transport in SSA. Results highlight the widespread disparities in accessibility between regions, using the policy relevant threshold of ensuring people live within 1 h of the nearest health facility.

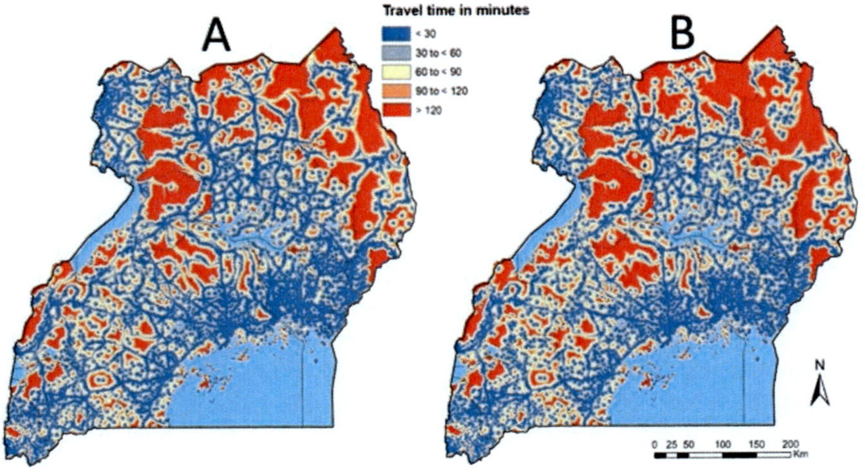

Fig. 6.3 Maps of Uganda showing travel time (in minutes) from each grid (100 × 100 m) to the nearest public health facility for (**a**) dry season and (**b**) wet season. They show increasing from 0 min (dark blue) to 100 min (dark red)

Table 6.2 Regional variation in geographic accessibility between the wet and dry seasons

	One hour		Two hours	
Region	Dry	Wet	Dry	Wet
Acholi	74.1	70.7	91.6	88.2
Ankole	81.1	79.0	98.2	98.0
Buganda	85.1	83.8	95.8	95.4
Bukedi	96.7	96.5	99.8	99.8
Bunyoro	77.4	71.5	96.6	95.6
Busoga	94.8	93.5	99.3	99.3
Elgon	96.4	96.3	99.6	99.5
Karamoja	64.6	61.1	87.7	86.5
Kigezi	91.9	91.3	98.7	98.7
Lango	82.9	78.4	99.0	98.7
Teso	85.4	84.7	99.2	99.2
Toro	72.5	71.2	94.2	94.0
West Nile	90.4	89.4	98.1	98.1
National average	84.1	82.1	96.8	96.2

Thus, the outputs can provide useful insights into where gaps in accessibility are, and decision makers can know where to narrow down to if service expansion is needed.

Results should however be interpreted in the context of some limitations. Other factors such as cost of transport, educational attainment, cultural factors and service acceptability are some of the factors that affect geographic access. Therefore including these variables is likely to provide a more complete picture of access to health services (Ouma et al. 2017). Traditionally, these factors have not been included in spatial accessibility models, but with the increasing availability of data at fine spatial resolution (Graetz et al. 2018), the possibilities of exploring multivariable spatial access models increase. Other factors rarely implemented in spatial access models include competition between facilities, especially in urban areas where patients are faced with multiple choices of where to attend. These are commonly implemented in the gravity models (Wan et al. 2012), but its reliance on accurate data on service capacity and population data at fine geographical units makes it untenable in African settings.

References

Aday, L. A., & Andersen, R. (1974). A framework for the study of access to medical care. *Health Services Research, 9*, 208–220.

Alegana, V. A., Wright, J. A., Petrina, U., Noor, A. M., Snow, R. W., & Atkinson, P. M. (2012). Spatial modelling of healthcare utilisation for treatment of fever in Namibia. *International Journal of Health Geographics, 11*, 6.

Alun, E. J., & David, R. P. (1984). *Accessibility and utilization: Geographical perspectives on health care delivery*. London: Harper & Row Ltd.

Austin, C. (2012). *Bike speed calculator*. Retrieved July 4, 2016 from http://bikecalculator.com/.

Blanford, J. I., Kumar, S., Luo, W., & MacEachren, A. M. (2012). It's a long, long walk: Accessibility to hospitals, maternity and integrated health centers in Niger. *International Journal of Health Geographics, 11*, 24.

Delamater, P. L. (2013). Spatial accessibility in suboptimally configured health care systems: A modified two-step floating catchment area (M2SFCA) metric. *Health Place, 24*, 30–43.

Delamater, P. L., Messina, J. P., Shortridge, A. M., & Grady, S. C. (2012). Measuring geographic access to health care: Raster and network-based methods. *International Journal of Health Geographics, 11*, 15.

Dixit, A., Lee, M.-C., Goettsch, B., Afrane, Y., Githeko, A. K., & Yan, G. (2016). Discovering the cost of care: Consumer, provider, and retailer surveys shed light on the determinants of malaria health-seeking behaviours. *Malaria Journal, 15*, 179.

Ensor, T., & Cooper, S. (2004). Overcoming barriers to health service access: influencing the demand side. *Health policy and planning 19*(2):69–79.

Evans, D. B., Hsu, J., & Boerma, T. (2013). Universal health coverage and universal access. *Bulletin of the World Health Organization, 91*, 10–11.

Feikin, D. R., Nguyen, L. M., Adazu, K., Ombok, M., Audi, A., Slutsker, L., & Lindblade, K. A. (2009). The impact of distance of residence from a peripheral health facility on pediatric health utilisation in rural western Kenya. *Tropical Medicine and International Health, 14*, 54–61.

Gautam, S., Li, Y., & Johnson, T. G. (2014). Do alternative spatial healthcare access measures tell the same story? *GeoJournal, 79*, 223–235.

Graetz, N., Friedman, J., Osgood-Zimmerman, A., et al. (2018). Mapping local variation in educational attainment across Africa. *Nature, 555*, 48–53.

Guagliardo, M. F. (2004). Spatial accessibility of primary care: Concepts, methods and challenges. *International Journal of Health Geographics, 3*, 3.

Haas, J. S., Phillips, K. A., Sonneborn, D., McCulloch, C. E., Baker, L. C., Kaplan, C. P., Perez-Stable, E. J., & Liang, S.-Y. (2004). Variation in access to health care for different racial/ethnic groups by the racial/ethnic composition of an individual's county of residence. *Medical Care, 42*, 707–714.

Higgs, G. (2004). A literature review of the use of GIS-based measures of access to health care services. *Health Services and Outcomes Research Methodology, 5*, 119–139.

Hu, R., Dong, S., Zhao, Y., Hu, H., & Li, Z. (2013). Assessing potential spatial accessibility of health services in rural China: A case study of Donghai County. *International Journal for Equity in Health*. https://doi.org/10.1186/1475-9276-12-35.

Huerta, M. U., & Källestål, C. C. (2012). Geographical accessibility and spatial coverage modeling of the primary health care network in the Western Province of Rwanda. *International Journal of Health Geographics, 11*, 40.

Lehner, B., & Döll, P. (2004). Development and validation of a global database of lakes, reservoirs and wetlands. *Journal of Hydrology, 296*, 1–22.

Levesque, J.-F., Harris, M. F., & Russell, G. (2013). Patient-centred access to health care: Conceptualising access at the interface of health systems and populations. *International Journal for Equity in Health, 12*, 18.

Linard, C., Gilbert, M., Snow, R. W., Noor, A. M., & Tatem, A. J. (2012). Population distribution, settlement patterns and accessibility across Africa in 2010. *PLoS One, 7*, e31743.

Luo, W. (2004). Using a GIS-based floating catchment method to assess areas with shortage of physicians. *Health Place, 10*, 1–11.

Macharia, P. M., Odera, P. A., Snow, R. W., & Noor, A. M. (2017a). Spatial models for the rational allocation of routinely distributed bed nets to public health facilities in Western Kenya. *Malaria Journal, 16*, 367.

Macharia, P. M., Ouma, P. O., Gogo, E. G., Snow, R. W., & Noor, A. M. (2017b). Spatial accessibility to basic public health services in South Sudan. *Geospatial Health, 12*, 510.

Maina, J., Ouma, P. O., Macharia, P. M., Alegana, V. A., Mitto, B., Fall, I. S., Noor, A. M., Snow, R. W., Okiro, E. A. (2019). A Spatial Database of Health Facilities Managed by the Public Health Sector in Sub Saharan Africa. *Scientific data 25*;6(1):134.

Makanga, P. T., Schuurman, N., Sacoor, C., et al. (2017). Seasonal variation in geographical access to maternal health services in regions of southern Mozambique. *International Journal of Health Geographics, 16*, 1.

Mao, L., & Nekorchuk, D. (2013). Measuring spatial accessibility to healthcare for populations with multiple transportation modes. *Health Place, 24*, 115–122.

Masoodi, M., & Rahimzadeh, M. (2015). Measuring access to urban health services using geographical information system (GIS): A case study of health service management in Bandar Abbas, Iran. *International Journal of Health Policy and Management, 4*, 439–445.

McGrail, M. R., & Humphreys, J. S. (2009). Measuring spatial accessibility to primary care in rural areas: Improving the effectiveness of the two-step floating catchment area method. *Applied Geography, 29*, 533–541.

MEASURE Evaluation. (2018). Master facility list. In *Health information systems strengthening*. Retrieved April 3, 2018 from https://www.measureevaluation.org/his-strengthening-resource-center/resources/master-facility-list.

Mennis, J. (2009). Dasymetric mapping for estimating population in small areas. *Geography Compass, 3*, 727–745.

Nesbitt, R. C., Gabrysch, S., Laub, A., et al. (2014). Methods to measure potential spatial access to delivery care in low- and middle-income countries: A case study in rural Ghana. *International Journal of Health Geographics, 13*, 25.

Neutens, T. (2015). Accessibility, equity and health care: Review and research directions for transport geographers. *Journal of Transport Geography, 43*, 14–27.

Ni, J., Wang, J., Rui, Y., Qian, T., & Wang, J. (2015). An enhanced variable two-step floating catchment area method for measuring spatial accessibility to residential care facilities in Nanjing. *International Journal of Environmental Research and Public Health, 12*, 14490–14504.

Noor, A. M., Gikandi, P. W., Hay, S. I., Muga, R. O., & Snow, R. W. (2004). Creating spatially defined databases for equitable health service planning in low-income countries: The example of Kenya. *Acta Tropica, 91*, 239–251.

Noor, A. M., Amin, A. A., Gething, P. W., Atkinson, P. M., Hay, S. I., & Snow, R. W. (2006). Modelling distances travelled to government health services in Kenya. *Tropical Medicine and International Health, 11*, 188–196.

Noor, A. M., Alegana, V. A., Gething, P. W., & Snow, R. W. (2009). A spatial national health facility database for public health sector planning in Kenya in 2008. *International Journal of Health Geographics, 8*, 13.

Okafor, U. V., Efetie, E. R., & Ekumankama, O. (2009). Eclampsia and seasonal variation in the tropics—A study in Nigeria. *The Pan African Medical Journal, 2*, 7.

Okwaraji, Y. B., Mulholland, K., Schellenberg, J. R., Andarge, G., Admassu, M., & Edmond, K. M. (2012). The association between travel time to health facilities and childhood vaccine coverage in rural Ethiopia. A community based cross sectional study. *BMC Public Health, 12*, 476.

Ouma, P. O., Agutu, N. O., Snow, R. W., & Noor, A. M. (2017). Univariate and multivariate spatial models of health facility utilisation for childhood fevers in an area on the coast of Kenya. *International Journal of Health Geographics, 16*, 34.

Ouma, P. O., Maina, J., Thuranira, P. N., Macharia, P. M., Alegana, V. A., English, M., Okiro, E. A., & Snow, R. W. (2018). Access to emergency hospital care provided by the public sector in sub-Saharan Africa in 2015: A geocoded inventory and spatial analysis. *The Lancet Global Health, 6*, e342–e350.

Owen, K. K., Obregón, E. J., & Jacobsen, K. H. (2010). A geographic analysis of access to health services in rural Guatemala. *International Health, 2*, 143–149.

Penchansky, R., & Thomas, J. W. (1981). The concept of access: Definition and relationship to consumer satisfaction. *Medical Care, 19*, 127–140.

Ray, N., & Ebener, S. (2008). AccessMod 3.0: Computing geographic coverage and accessibility to health care services using anisotropic movement of patients. *International Journal of Health Geographics, 7*, 63.

RCMRD. (2017). *Uganda sentinel 2 land use land cover*. Retrieved June 17, 2018 from http://geoportal.rcmrd.org/layers/servir%3Auganda_sentinel2_lulc2016.

Rodrigue, J.-P., Comtois, C., & Slack, B. (2013). *The geography of transport systems* (3rd ed.). New York: Routledge.

Rose-Wood, A., Heard, N., Thermidor, R., et al. (2014). Development and use of a master health facility list: Haiti' s experience during the 2010 earthquake response. *Global Health: Science and Practice, 2*, 357–365.

Rutherford, M. E., Dockerty, J. D., Jasseh, M., et al. (2009). Access to health care and mortality of children under 5 years of age in the Gambia: A case-control study. *Bulletin of the World Health Organization, 87*, 216–224.

Rutherford, M. E., Mulholland, K., & Hill, P. C. (2010). How access to health care relates to under-five mortality in sub-Saharan Africa: Systematic review. *Tropical Medicine and International Health, 15*, 508–519.

Schoeps, A., Gabrysch, S., Niamba, L., Sié, A., & Becher, H. (2011). The effect of distance to health-care facilities on childhood mortality in rural Burkina Faso. *American Journal of Epidemiology, 173*, 492–498.

Schuurman, N., Bérubé, M., & Crooks V a. (2010). Measuring potential spatial access to primary health care physicians using a modified gravity model. *The Canadian Geographer/Le Géographe canadien, 54*, 29–45.

Tansley, G., Schuurman, N., Amram, O., & Yanchar, N. (2015). Spatial access to emergency services in low- and middle-income countries: A GIS-based analysis. *PLoS One, 10*, e0141113.

Tobler, W. (1993). *Three presentations on geographical analysis and modeling: Non-isotropic geographic modeling; Speculations on the geometry of geography*. Santa Barbara: Global Spatial Analysis.

USAID, WHO. (2018). *Master facility list resource package*.

Vora, K. S., Yasobant, S., Patel, A., Upadhyay, A., & Mavalankar, D. V. (2015). Has Chiranjeevi Yojana changed the geographic availability of free comprehensive emergency obstetric care services in Gujarat, India? *Global Health Action, 8*, 10–11.

Wan, N., Zou, B., & Sternberg, T. (2012). A three-step floating catchment area method for analyzing spatial access to health services. *International Journal of Geographical Information Science, 26*, 1073–1089.

Wang, F. (2012). Measurement, optimization, and impact of health care accessibility: A methodological review. *Annals of the American Association of Geographers, 102*, 1104–1112.

WHO. (2010). *Monitoring the building blocks of health systems: A handbook of indicators and their measurement strategies*. Geneva: Switzerland.

WHO. (2012). *Creating a master healthy facility list, draft*. Geneva: Switzerland.

Zorn, A. (2008). *Bicycle velocity and power calculator*. Retrieved July 4, 2016 from http://www.kreuzotter.de/english/espeed.htm.

Part II
Health and Place

Chapter 7
Representing Health: An Afrocentric Perspective from Ghana

Amos Laar, John Ganle, Adobea Owusu, Eric Tenkorang, Yetsa A. Tuakli-Wosornu, Ireneous Soyiri, Michael Okyerefo, and Kodjo Senah

Abstract Health and ill health are critical concepts in public health with far reaching socio-political implications. Notwithstanding, conceptualizing health has been challenging as there appears to be no definitional consensus. A related challenge is finding a sociologically, politically, spiritually, and culturally-sound cleavage between health as defined, and health as experienced in different settings. Scholars have attributed the lack of consensus to the complex and interdisciplinary nature of health. In this chapter, we build on the classic definition of health by the World Health Organization (WHO) to provide an Afrocentric perspective of the concept. While there have been several attempts to move away from the dominant biomedical foundations of health, an Afrocentric perspective is currently lacking in the health representation literature. We fill this important gap by using case studies that emphasize the African traditional notions of good health and illustrate how health is more than just the absence of disease in the living. The chapter begins with a review of concepts of health, then continues to provide a nuanced analysis of how both

A. Laar (✉) · J. Ganle
Department of Population, Family and Reproductive Health, School of Public Health, University of Ghana, Legon, Accra, Ghana
e-mail: alaar@ug.edu.gh

A. Owusu
Institute of Statistical, Social and Economic Research (ISSER), University of Ghana, Legon, Accra, Ghana

E. Tenkorang
Department of Sociology, Memorial University of Newfoundland, St. John's, NL, Canada

Y. A. Tuakli-Wosornu
Department of Chronic Disease Epidemiology, Yale School of Public Health, New Haven, CT, USA

I. Soyiri
Usher Institute of Population Health Sciences and Informatics, The University of Edinburgh, Edinburgh, Scotland, UK

M. Okyerefo · K. Senah
Department of Sociology, University of Ghana, Legon, Accra, Ghana

P. T. Makanga (ed.), *Practicing Health Geography*, Global Perspectives on Health Geography, https://doi.org/10.1007/978-3-030-63471-1_7

medical and indigenous African perspectives of health are important in the global representation of health.

Keywords Representing health · Health · Disease · Indigenous health · Global health

Introduction

Health: Definitions and Historical Perspectives

Around the world, good health has historically been viewed as a supernatural gift from the gods, and disease, a punishment from the same source. As a result, prayers and sacrifices were/are often offered to the gods for 'good health', in the quest for healing, and to spare individuals and populations of the misfortune of 'bad health' (Sigerist 1961). Medical and philosophical texts from the sixth to fourth centuries BC, confirm that the ancient Greeks were the first to break away from supernatural, non-biologic conceptions of health and disease (Tountas 2009). Transitioning from health, the supernatural, to health, the interaction between human beings and the environment, the ancient Greeks developed the physiocratic school of thought, realizing that:

1. Maintaining good health and fighting illness depends on natural causes
2. Health and disease cannot be dissociated from physical and social environments nor from human behaviour
3. Health may be best defined as a state of dynamic equilibrium between the internal and the external environments (Yapijakis 2009)

This evolution in the understanding of health was crucial to realizing that individuals and their physical and social environments were important for achieving optimum health. Man's understanding of health has since undergone a transformative metamorphosis. Following the foundational work of Hippocrates, Claudius Galenus, or Galen (130–201 AD) further developed the theory and practice, and carried Greco-Roman medicine to its zenith (Ergil et al. 1997). In the early 160s AD, Galen postulated that a physician needed to study the entire body (rather than a single organ system) to get a complete understanding of patients' health (Ergil et al. 1997). Thus, the holistic idea of health that considers the whole being, physical, mental, and emotional was born. The miasma theory of health, the idea that bad air 'miasma' was the cause of disease, became popular from first century BC to the middle ages/1880 (Bloom 1965; Sterner 1948). Then came the spontaneous generation theory, which suggested that living things arose from non-living matter (Brack 1998). Undoubtedly, the work of Francesco Redi, Louis Pasteur and Robert Koch clarified the existence of germs, thus the Germ Theory era began (Tyndall 1905; Karamanou et al. 2012). From holistic wellness to Germ Theory, the evolution of health has been significant. The definition of health has changed from its supernatural and metaphysical foundations to a broader but idealistic conceptualization as captured by the WHO. The WHO defines health as a state of complete physical, mental and social well-being and not merely the absence of disease or infirmity (WHO 1948).

Alongside the WHO's, several definitions of health exist. None has so far been able to give a universally acceptable, yet culturally-sensitive appeal to all (Boddington and Räisänen 2009). Following a reading of such definitions by many (Yapijakis 2009; Patrick et al. 1973; Bircher 2005), we confirm that each one of them fits into Wylie's dichotomous concepts of an asymptotic/open-ended or an elastic model of health (Wylie 1970). In the open-ended type, health is an ideal that can never be fully attained and death occurs just below the absolute zero of the scale which has no fixed limit for maximum human functioning. The WHO definition, according to Wylie, fits into the open-ended model. The elastic type on the other hand defines health as the ability to resist threats of disease and the positive interaction between a person and the environment or community, Wylie notes.

In this chapter, we discuss indigenous African conceptions of health and illness, which are deeply rooted in the socio-cultural, political, and spiritual realms. We initiate a discussion on how to integrate these views into a more contextualized and acceptable global representations of health.

Indigenous Concepts of Health (Africa)

Within many indigenous African beliefs, conceptions of health and illness are deeply rooted in the spiritual realm (Asare and Danquah 2017; University of Massachusetts 2017), with the belief that what is seen in the physical realm is inherently controlled by the spiritual (Asare and Danquah 2017).

In Africa, people tend to see health as the ability to maintain specific functions, i.e. working and fulfilling societal roles, contributing to the home and fulfilling familial roles, and/or worshipping and contributing to the community's religious or spiritual wellness (Gessert et al. 2015). But traditional African concepts of health also go beyond the 'living', pivoting around family members, community, and ancestors in the other world. This integrated view of health is based on the African view of reality (Omonzejele 2008) where intrarelationships are just as important as interrelationships. There are relationships within oneself, relationships between the physical, the living; and relationships with the spiritual (ancestors) as well.

It is of paramount importance that the ancestors are also kept healthy so that they can protect the living. Among many urban and traditional communities in West Africa (i.e. Nigeria, Ghana, Benin, etc.), it is customary to offer sacrifices to the ancestors and deities. For instance, Omonzejele (2008) notes that amongst the Bini and Esan people of Nigeria, placing parts of their meals/drinks in certain corners of their homes, designated as ancestral shrines, is a common practice by families. And this is done to enable them to perform their ancestral functions, which are principally for the welfare of the living. Further, life in Africa is cyclical due to the belief that the ancestors of today could reincarnate as the living of tomorrow (Omonzejele 2008).

Although African indigenous concepts of health and illness do not exclude Western medicine and its definitions, it is believed that specific diseases are only curable by indigenous means (Oti-Boadi 2017; Okyerefo and Fiaveh 2017). For example, in Ghana, suffering, illness, and death are often attributed to witchcraft (University of Massachusetts 2017), especially in the face of challenging diagnoses like mental illness (Danquah 1982; Mantey 2010; Asare and Danquah 2017).

In the African culture, illness is also believed to be a private and personal affair, discussed secretly among close family members or sometimes even a source of disgrace to the sick person and the immediate family (University of Massachusetts 2017). This results from the perception that illness is a sign of weakness, often induced by a curse and invoked by others due to maleficence or as a result of sin committed against God or deities or other lesser spirits (Danquah 1982; Mantey 2010). The secrecy around illness surfaces in social settings: it is culturally inappropriate to ask what caused someone's demise, particularly when the inquiry does not come from a close family member.

In their study on prayer and health-seeking beliefs in Ghana, Okyerefo and Fiaveh (2017) discovered among prayer group members that, disease has spiritual and physical origins, which is why they would seek (when they are sick), both biomedical and spiritual care. The prayer group members studies 'were convinced that hospitals can help with physical diseases but spiritual diseases require spiritual solutions, such as prayer' (Okyerefo and Fiaveh 2017: 1). For example, they would differentiate between physical malaria and spiritual malaria. Physical malaria they know to be caused by the mosquito while they believe witches can cause spiritual malaria, just as other ailments. Medical doctors can, thus, treat physical malaria, but only God can heal spiritual malaria, which is malevolent and recurring. Given the rising numbers of such prayer group members in Africa—precipitated partly by the upsurge in Pentecostal-charismatic Christianity, their beliefs and cultural practices form part of those of a huge constituency in the field of healthcare in Africa.

This deeply-rooted spiritual concept of health and illness need not be condemned. It gives hope to indigenous African believers (Okyerefo and Fiaveh 2017; Asare and Danquah 2017; Oti-Boadi 2017), and can facilitate healing and recovery when combined with Western conceptions and treatment practices (Plante 2005; Asare and Danquah 2017; Okyerefo and Fiaveh 2017; Oti-Boadi 2017). Such conceptions are inherently and inseparably African, resonate with the story of God's creation (Asare and Danquah 2017), and are additionally rooted in the belief in the existence and power of other supernatural spirits and beings (Kiev 1964; Mbiti 1970; Omonzejele 2008; Danquah 2014; Appiah et al. 2007; Mantey 2010; Okyerefo and Fiaveh 2017). They are often passed on from generation to generation (Asare and Danquah 2017).

Asare and Danquah (2017: 49) have labelled this generational transfer of beliefs, and its influence on thinking and behaviour, including the conceptualization of health, well-being, and illness through this dogma, as 'behaviour genetics'. Even for the elite Africans who are well-versed in the biomedical/Western/biopsychosocial model of health, illness and health typically operate within a duality, with a spiritual current running alongside the biopsychosocial model (University of Massachusetts 2017).

In contrast to the indigenous model of health and illness, the biomedical/biopsychosocial model operates with the view of health as a combined element of mental, emotional, social and physical factors (Engel 1980; Asare and Danquah 2017), and explains health and illness using the germ theory of disease (Danquah 1982; Mantey 2010). As Asare and Danquah (2017: 49) write:

> For the African, however, wellbeing is not just about the healthy functioning of the body system through proper healthcare and lifestyle, but well-being goes beyond scientific causes to include spiritual involvement (also see Kiev 1964; Mbiti 1970); which constitutes [a] "modification of the biopsychosocial model to include spiritual factors" (McKee and Chappel 1992; Asare and Danquah 2017: 49)

Case Studies

In Ghana and Africa as a whole, the following two case studies give ample support to the views of previous authors as articulated in the literature. Case 1 is on primary research with data collected in April and May 2018. Informed consent was sought from the family of the first case. Case 2 drives from an earlier research that aimed to appreciate local African concepts health and causal knowledge on malaria.

Case 1. Mental Illness

In a 'Garden' (healing yard) of a faith-healing, syncretic[1] *church based in the Eastern Region of Ghana, a woman who is suspected to have a 'spiritual' illness is sent to a renowned Prophetess/healer within the locality for help. The family of the sick woman did not consider sending her to an orthodox hospital in the first place. This is explained mostly by the family's view that orthodox healthcare centres may not have the full-fledge expertise and related resources for care, and also by inadequate family resources for paying the fees for orthodox centres which happen to be more expensive and will require on-the-spot payment. Further, the family felt that such orthodox healthcare centre would mostly not resolve the suspected spiritual undertone of the disease being presented.*

Akosua Fio (a pseudonym) is 42-years-old, divorcee, has four children–mostly teenagers, did not complete basic schooling and has abandoned her vocation as a seamstress cum *baker. She has lived in the capital city of Accra for more than 10 years trying to eke a living. Her children are growing, and given her quite affluent parental background, feels that the children should attend the best and competitive*

[1] Twumasi (2005) characterizes syncretic churches as those that combine both indigenous and orthodox modes of a supernatural being and worship.

secondary and tertiary schools. The children's father barely supports them. Akosua's mother herself has been maimed for nearly three decades, a disease that happened overnight, a swollen leg and thigh, which has defied treatment and left her home-bound for all these years; suspected to be a spiritually caused because it happened overnight after she has visited a farm where her larger extended family is making claims to a portion of.

Akosua was sent to the 'Garden' because she has of late, repeatedly been running intermittently into the streets to shout and talk, mentioning that she feels there is pepper in her head, and all over her body. She has had sleepless nights for several months, and worries deeply about the well-being of her children. She is also very angry with the behaviour of her siblings towards her. Akosua accuses her mother of being a witch, for which she consulted a 'Mallam' (usually a Muslim religious leader cum *combined indigenous healer) who informed her after divination that her plight is spiritual, and caused by her mother, through witchcraft. She in turn asked the Mallam to 'arrest and main' her mother so she could not continue coming after her in the spiritual realm to hurt her.*

The syncretic Odiyifoo (Prophet/Prophetess, in Akan) diagnosed Akosua as follows: (1) her honour has been spiritually usurped by her sisters, especially the older one who is the only sibling with a (rich) husband among them; (2) she has been sold spiritually. Resultantly, the Odiyifoo prescribed the following solutions: (a) An offering should be made with a fowl and a specified amount of money (changed into small denominations, including coins) thrown by a riverside. (b) The offering should include cooking plenty rice and stew with chicken (a favourite Ghanaian, Western-originated festive meal) and inviting the congregation at the syncretic church to eat after a Friday forenoon worship which is their special 'prayer and deliverance session'. (c) A lot of children should be invited to partake in the meal. This will be a form of appeasing persons in the spiritual realm so her children would also be saved from the predicament her sickness is supposed to usher unto them. (d) Akosua's head is shaved, and white kaolin is smeared around her head two times daily. (e) She is instructed to put 11 buckets of water under the cross at the centre of the 'Garden'. (f) To carry a big bowl of water on her head and walk around the cross seven times each day. (g) To bath with the 11 buckets of water daily for 14 days. (h) Akosua has to fast for 3 weeks (21 days), for at least half a day each successive day.

Carrying the bowl of water to go round the cross is meant to 'bring down her attackers and any evil spirit working against her as happened in the story of the Wall of Jericho in the Bible' (Joshua 6: 1–27), and bathing the 11 buckets for 14 days is supposed to drive away any evil omen that comes her way. Seven is an important biblical number,[2] *and having multiples of seven as part of the healing is in tune with scripture, as the Odiyifoo described. Eleven is also a fantastic number*

[2]Theologically, the number seven represents both spiritual and physical completeness and perfection. Source: http://www.biblestudy.org/bibleref/meaning-of-numbers-in-bible/7.html. Accessed May 12, 2018.

to work with spiritually, because it represents the 12 disciples of Jesus, without 'evil and traitor' Judas.

Case 2. Parasitic Disease

This case study discusses local malaria concepts and causal knowledge. The case is based on qualitative research conducted among local community members (children and adults aged 10–65 years) and healthcare professionals across urban and rural areas. The study aimed to understand local conceptions of malaria and the implications for public health preventive and biomedical treatment interventions (Ganle 2012).

In biomedicine, malaria is recognized as a preventable and treatable infection in sub-Saharan Africa, spread by the female Anopheles mosquito. General awareness of malaria in sub-Saharan Africa is almost universal, albeit with local appreciations and interpretations (Agyepong 1992; Kengeya-Kayondo et al. 1994; Elzubier et al. 1997; Brieger et al. *2001; Comoro et al. 2003). In biomedical literature on malaria however, these local interpretations are often termed "misconceptions" (Agyepong 1992; Elzubier et al. 1997). There are however different conceptions around malaria. In several studies conducted in Africa on malaria, health professionals and educated locals explain malaria and its symptoms in ways that corresponded to standard biomedical/clinical definitions (Beiersmann et al. 2007; Williams and Jones 2004; O'Neill et al. 2015; Gessler et al. 1995; Ganle 2012). In one of these studies, malaria was presented by health experts as a natural disease, and its cause and transmission were framed within the germ theory of disease: that a specific disease results from a specific causal micro-organism. The female Anopheles mosquito was often identified as the vector transmitting malaria (Ganle 2012). While these health professionals and some educated locals spoke of malaria in naturalistic and mono-causal terms, many locals spoke about malaria as an ambiguous and contestable disease. Generally, there is no exact terminology or local illness concept that approximates the biomedical definition of malaria. Rather, various local terminologies are used to denote the disease including donne baalong (mosquito disease), malareya and feba. These conceptions are consistent with findings among the Dangme of Ghana, where asra, (local term for malaria) was found to be a highly contestable illness concept as it could also be attributed to other illness conditions. One of the studies indicated that, local explanatory models for malaria were subsumed under diverse medical cosmologies. Several participants noted that mosquitoes alone could not cause malaria:*

> *Mosquitoes? Me I don't think only this thing…mosquito, causes the malareya! I believe there are others that the doctors don't know or don't want to tell us.*
>
> *See… mosquitoes bite everyone in this village, but some don't get the feba… sometimes those who constantly sleep under mosquito net even get feba…why? I think there must be other causes.*
>
> *You see, bad things don't just happen to people without cause…A disease like malaria may be the result of bad luck or disequilibrium in social relations or moral infractions rather than simple mosquito bites.*

Discussion

The case studies illustrate deeply-rooted political, social, cultural, and spiritual beliefs of the African (Mbiti 1970; Danquah 1982; Mantey 2010; Asare and Danquah 2017; Okyerefo and Fiaveh 2017; Ganle 2012), which influence notions of good and ill health, causation of morbidity, and requirements for disease treatment. In Africa, Western-based healthcare centres are not the first places families look to for healthcare (Asare and Danquah 2017; University of Massachusetts 2017). Healthcare-seeking behaviour in many parts of Africa is underlaid by indigenous beliefs, which are informed by intergenerational cultural experiences (Asare and Danquah 2017). Often, disease causation and presentations that are difficult to explain, are assigned a spiritual explanation.

Despite Africa's long exposure to orthodox Western healthcare, it is still commonplace to rely on indigenous healthcare (Osei 2004; Iroegbu 2005; Asare and Danquah 2017; Danquah 1982; Mantey 2010), due to the desire for holistic healing (Gyasi et al. 2016; Okyerefo and Fiaveh 2017) and also the perceived inadequacies of the biomedical and biopsychosocial healthcare models (Danquah 1982; Mantey 2010; Gyasi et al. 2016; Okyerefo and Fiaveh 2017; University of Massachusetts 2017). Of note, cost concerns are not typically brought to the fore during healthcare-seeking decision making. The need to satisfy the perceived spiritual and socio-cultural underlying reason for ill-health takes primary importance (Danquah 1982; Mantey 2010; Okyerefo and Fiaveh 2017).

The case studies also show that based on the perceived aetiology of the sickness, indigenous African healthcare practitioners may combine different modes of healing, such as herbal, prayer and fasting, and exorcism of spirits (Danquah 1982; Osei 2004; Mantey 2010; Asare and Danquah 2017; Okyerefo and Fiaveh 2017). Often, herbal treatment is the first choice, followed by spiritual treatment, and finally, Western biomedical treatment (Asare and Danquah 2017). The spiritual nature of health is further emphasized in Case 2. In this case, ill-health and medicine as concepts that are located in diverse realms, including astrology, alchemy, superstition, magic, divination, and modernity. These medical traditions and beliefs about health and disease offer language, ideas, expectations, cognitive models, and normative guidelines through which people speak about their experience of health and disease, popular notions of illness causation, and preventive models. In the context where this case study emerged, illness is identified in ways similar to the etiological categories identified by Feierman (1981) in Tanzania. People first spoke of "*malareya* of God or nature" as opposed to *malareya* caused by humans or spirits. "*Malareya* of God" comes closest to the "natural" or germ theory explanation of malaria as it relates to general principles of the behaviour of nature other than artificial human intervention. Mosquitoes were identified as the cause of this first category and people spoke of it as happening by accident or negligence. The second is *malareya* from sorcery or witchcraft, which comes with ill-intentioned humans as the architects. Finally, people spoke of *malareya* from moral infractions and the neglect of ancestral spirits (*Kpime*). In other words, malaria could be the result of "sinful" or morally wrong or unjustified behaviour, witchcraft, and punishment from gods and ancestral spirits. Malaria is thus not a natural, objectively neutral disease, but has moral undertones. Indeed, many of the depictions of malaria in this study have been reported in other parts of Africa, including Ghana (Agyepong 1992), Tanzania (Comoro et al. 2003), Uganda (Kengeya-Kayondo et al. 1994), and Nigeria (Brieger et al. 2001).

While such lay discourses certainly contain mystical ideas about malaria, which may prove problematic when it comes to instituting prevention interventions, they nevertheless challenge the dominant biomedical meta-narratives, which speak of malaria in naturalistic and mono-causal terms. Indeed, this case study suggests that rather than viewing such ambiguities as "misconceptions" or the "exception", the "deviate", the "parochial", and the "traditional" in the face of the "modern", they ought to be seen as constitutive of the diverse cosmic life-worlds in which people lead their lives.

Orthodox Western clinicians in African settings are thus encouraged to empathize with the spiritual beliefs of African patients, and leverage such beliefs for medical compliance, adherence and recovery. The contribution of African spiritual healers should also be acknowledged and shaped positively to the advantage of patients and society in general. For instance, were traditional healers trained in basic medical conditions, they may be able to modify their practice and partner with Western healers to the advantage of society. Orthodox healthcare practitioners and other social workers could also review the spiritual beliefs of patients and their families to more readily pinpoint ambivalence about disease aetiology, and potentially challenge misconceptions and mistrust, to help facilitate the healthcare provision. In sum, the multiple ways in which health or ill health is understood in Africa (as per these cases), challenges the current seemingly unitary/closed representation of health by the dominant models. Our argument thus has been that, the naturalness with which the dominant models of health ignore the variegated, pluralistic and dynamic nature of health—per the contemporary African perspectives presented, does not auger well for global health.

Reconceptualizing, Reframing, and Representing Health

For nearly a century, humanity (expert and lay) has been compelled (explicitly or implicitly) to use concepts of health propounded by Western experts. For instance, the dominant, 70-year-old WHO definition of health, authored by an expert panel, is broad and aims at attaining the highest possible level of health (WHO 1948). Its proponents argue that the definition expanded the concept of health from mere concern about the physical characteristics of a disease to a consideration of the social determinants which affect a patient's health outcomes and quality of life. The underlying problem with the definition has been its impracticality, opponents note. While the definition is positive, ambitious, and offers unlimited opportunities for global improvement, it is not practical because it is too fundamental and cannot be reliably and equitably enforced (Badash et al. 2017; Doll 1992; Habersack and Luschin 2013).

The definition has also been criticized for its conceptual drawbacks (Saracci 1997). We fail to see how this concept of health accommodates the socio-spiritual realities of traditional African understandings of health. To be comprehensive and contextually-sensitive, any definition of health must respect the fundamental health needs, existing health concepts and experiences of the end-user. Such a definition should highlight setting-driven differences. That the social is important in the medical is not a new idea. In the African setting, socioenvironmental, socio-cultural, socio-spiritual dimensions are intimately interwoven in all diseases. As discussed within this chapter, socio-spiritual drivers and contexts influence health course, severity, and quality of life (Mantey 2010; Asare and Danquah 2017; Okyerefo and Fiaveh 2017). Even congenital and degenerative conditions, are subject to the influences of social context in how they are managed. Allotey et al. (2003) argue that clinically similar cases of ill health translate into different consequences, depending on individual circumstances at the micro, meso, and macro levels of context, where the contrast of disease consequences is especially evident between high-income and low-income countries.

We argue that, any attempt at arriving at an appropriate definition of health should be a deliberative and inclusive process. Health experts should have a say, and the lay persons too. We argue that such a definition of health should incorporate values that bear on ethics, culture, social frameworks, and spirituality. In this era where value-based medicine is taking off, such a definition would be welcome. While agreeing that the 1948 WHO definition of health is valid in 2018, we suggest a conceptual extension. Thus, if socio-cultural factors drive the prevalence of ill health, and socio-spiritual factors drive people's lived experiences, consequences, and course of the ill health, then those factors ought to be considered in any conception of health. Such conceptualization of health would be useful for both lay and expert, and flexible enough to accommodate the heterogeneity in the notations of health. By emphasizing patients' values, their lived experiences, and concepts, the healthcare system would be better prepared to satisfy patients' health needs.

Reconceptualizing health in ways that acknowledge diverse and deep-seated cultural nuances, may help to fill-out the current (and arguably, incomplete) representation. Redefining health is an ambitious and complex process. Many factors need to be considered, many stakeholders consulted/engaged, and many cultures must be reflected. Such an exercise should seek to coalesce the values that underlie diverse definitional perspectives of health. The end result, and indeed the process itself, may improve clinicians' health literacy and cultural competencies, thereby optimizing care encounters in Africa and beyond.

References

Agyepong, I. A. (1992). Malaria: Ethnomedical perceptions and practice in an Adangbe farming community and implications for control. *Social Science & Medicine, 35*, 131–137.

Allotey, P., Reidpath, D., Kouamé, A., & Cummins, R. (2003). The DALY, context and the determinants of the severity of disease: An exploratory comparison of paraplegia in Australia and Cameroon. *Social Science & Medicine, 57*, 949–958.

Appiah, P., Appiah, K. A., & Agyeman-Duah, I. (2007). *Proverbs of the Akans: Bu Me Bε* (p. 336). Oxfordshire: Ayebia Clarke Publishing Ltd.

Asare, M., & Danquah, S. A. (2017). The African belief system and the patient's choice of treatment from existing health models: The case of Ghana. *Acta Psychopathology, 3*(4), 49.

Badash, I., Kleinman, N. P., Barr, S., Jang, J., Rahman, S., & Wu, B. W. (2017). Redefining health: The evolution of health ideas from antiquity to the era of value-based care. *Cureus, 9*(2), e1018. https://doi.org/10.7759/cureus.1018

Bircher, J. (2005). Towards a dynamic definition of health and disease. *Medicine, Health Care and Philosophy, 8*(3), 335–341.

Bloom, B. L. (1965). The "medical model," miasma theory, and community mental health. *Community Mental Health Journal, 1*(4), 333–338.

Boddington, P., & Räisänen, U. (2009). Theoretical and practical issues in the definition of health: Insights from Aboriginal Australia. *Journal of Medicine and Philosophy, 34*(1), 49–67.

Brack, A. (1998). Introduction. In A. Brack (Ed.), *The molecular origins of life* (p. 1). Cambridge: Cambridge University Press, ISBN: 978-0-521-56475-5.

Brieger, W. R., Sessay, H. R., Adesina, H., Mosanya, M. E., Ogunlade, P. B., Ayodele, J. O., & Arisasona, S. A. (2001). Urban malaria treatment behaviour in the context of low transmission in Lagos, Nigeria. *African Journal of Medical Science, 30*, 7–15.

Comoro, C., Nsimba, S. E. D., Warsame, M., & Tomson, G. (2003). Local understanding, perceptions and reported practices of mothers/guardian and health workers on childhood malaria in a Tanzania District—Implications for malaria control. *Acta Tropica, 87*, 305–313.

Danquah, S. A. (1982). The practice of behavior therapy in West Africa: The case of Ghana. *Journal of Behavior Therapy and Experimental Psychiatry, 13*(1), 5–13.

Danquah, S. A. (2014). *Clinical psychology in Ghana: History, training, research and practice*. Saarbrucken: Lambert Academic Publishing.

Doll, R. (1992). Health and the environment in the 1990s. *American Journal of Public Health, 82*, 933–941.

Engel, G. L. (1980). The clinical application of the biopsychosocial model. *The American Journal of Psychiatry, 137*, 535–544.

Ergil, K. V., Ergfl, M. C., Furst, P. T., Gordon, N., Janzen, J. M., Sobo, E. J., & Sparrowe, L. (1997). *Ancient healing-unlocking the mysteries of health and healing through the ages* (pp. 68–95). Lincolnwood, IL: Publications International Ltd., ISBN-10: 785324313.

Feierman, S. (1981). Therapy as a system-in-action in northeastern Tanzania. *Social Science & Medicine. Part B: Medical Anthropology, 15*(3), 353–360.

Ganle, J. K. (2012). "Today we have development but mosquitoes are everywhere": Development, Environmental Change and Malaria in a Rural District of Ghana. Human Welfare, 1:71–90.

Gessert, C., Waring, S., Bailey-Davis, L., Conway, P., Roberts, M., & VanWormer, J. (2015). Rural definition of health: A systematic literature review. *BMC Public Health, 15*(1), 378.

Gyasi, R. M., Asante, F., Abass, K., Yeboah, J. Y., Adu-Gyamfi, S., & Amoah, P. A. (2016). Do health beliefs explain traditional medical therapies utilisation? Evidence from Ghana. *Cogent Social Sciences, 2*(1), 1209995.

Habersack, M., & Luschin, G. (2013). WHO-definition of health must be enforced by national law: A debate. *BMC Medicine, 14*, 24.

Iroegbu, P. (2005). Beginning, purpose and end of life. Kpim of morality ethics, 440–445.

Karamanou, M., Panayiotakopoulos, G., Tsoucalas, G., Kousoulis, A. A., & Androutsos, G. (2012). From miasmas to germs: a historical approach to theories of infectious disease transmission. *Le Infezioni Medicina, 20*(1), 58–62.

Kengeya-Kayondo, J. F., Seely, J. A., Kajura-Bejenja, E., Kabunga, E., Mubiru, E., Sembaija, F., & Mulder, W. D. (1994). Recognition, treatment seeking behaviour and perception of cause of malaria among rural women in Uganda. *Acta Tropica, 58*, 267–273.

Kiev, A. (1964). *Magic, faith and health*. New York: The Free Press.

Mantey, J. (2010). *Traditional beliefs play major role in Ghanaian health care*. Retrieved May 11, 2018 from https://www.voanews.com/a/ghana-traditional-health-care-voa-80288697/416798.html.

Mbiti, J. S. (1970). *Concepts of god in Africa* (Vol. 348). London: The Camelot Press Ltd.

McKee, D. D., & Chappel, J. N. (1992). Spirituality and medical practice. *The Journal of Family Practice, 35*, 205–208.

Okyerefo, M. P. K., & Fiaveh, D. Y. (2017). Prayer and health- seeking beliefs in Ghana: Understanding the 'religious space' of the urban forest. *Health Sociology Review, 26*(3), 308–320. https://doi.org/10.1080/14461242.2016.1257360.

Omonzejele, P. F. (2008). African concepts of health, disease, and treatment: An ethical inquiry. *Explore: The Journal of Science and Healing, 4*(2), 120–126.

Osei, A. O. (2004). Types of psychiatric illnesses at traditional healing centres in Ghana. *Ghana Medical Journal, 35*, 106–110.

Oti-Boadi, M. (2017). Exploring the lived experiences of mothers of children with intellectual disability in Ghana. *SAGE Open, 7*(4), 2158244017745578.

Patrick, D. L., Bush, J. W., & Chen, M. M. (1973). Toward an operational definition of health. *Journal of Health and Social Behavior, 14*, 6–23.

Plante, T. G. (2005). *Contemporary clinical psychology* (2nd ed., p. 618). New York: John Wiley & Sons.

Saracci, R. (1997). The World Health Organisation needs to reconsider its definition of health. *British Medical Journal, 314*, 1409–1410. Retrieved from https://www.ncbi.nlm.nih.gov/pmc/articles/PMC2126653/.

Sigerist, H. E. (1961). *A history of medicine* (Early Greek, Hindu and Persian medicine) (Vol. II). New York: Oxford University Press.

Sterner, C. S. (1948). A brief history of miasmic theory. *Bulletin of the History of Medicine, 1948*(22), 747.

Tountas, Y. (2009). The historical origins of the basic concepts of health promotion and education: The role of ancient Greek philosophy and medicine. *Health Promotion International, 24*(2), 185–192.

Twumasi, P. A. (2005). *Medical systems in Ghana: A study in medical sociology*. Tema: Ghana Publishing Corporation.

Tyndall, J. (1905). *Fragments of science* (Vol. 2). New York: P. F. Collier, Chapters IV, XII (1876), XIII (1878). Retrieved from https://archive.org/details/fragmenoscien02tyndrich.

University of Massachusetts. (2017). *Cultural approaches to pediatric palliative care in central Massachusetts: Ghanaian*. Retrieved May 11, 2018 from https://libraryguides.umassmed.edu/c.php?g=499760&p=3422585.

WHO. (1948). Constitution of the World Health Organization. Basic Documents, Geneva, Switzerland.

Wylie, C. M. (1970). The definition and measurement of health and disease. *Public Health Reports, 85*(2), 100.

Yapijakis, C. (2009). Hippocrates of Kos, the father of clinical medicine, and Asclepiades of Bithynia, the father of molecular medicine. *In Vivo, 23*(4), 507–514.

Chapter 8
Access to Maternal Health in Regions of Rwanda: A Qualitative Study

Germaine Tuyisenge

Abstract Access to health care is essential to achieving the best health outcomes. In this chapter, I highlight the structural, geographical, social-cultural and economic factors that act as facilitators or barriers to accessing maternal health services at the community level in Rwanda. Using five dimensions of access to health care (availability, accessibility, accommodation, affordability and acceptability), I qualitatively explore community dynamics pertaining to women's access to community maternal health services in both rural and urban settings. I also examine the role of volunteer community health workers, who are elected by their own communities to provide maternal health services to women, and link communities to the formal health care system.

Keywords Access · Barriers · Facilitators · Maternal health · Health care · Community health workers · Rwanda

Introducing Access to Health Care

For decades, researchers and policy makers have tried to conceptualize the notion of 'access' as it relates to health care services in order to better understand the ways in which access (or inaccessibility) shapes health outcomes. Even though much progress has been made, there is no single conventional definition of access to health care. Instead, many definitions have been provided that reflect varying local contexts and disciplinary concepts. Khan and Bhardwaj (1994) point out that this has led to vastly different understandings of what it means to have access operating simultaneously. A common approach to understanding access involves conceptualizing factors that contribute to access and the ways in which they interact to shape different outcomes. Penchansky and Thomas's widely-cited concept of access, for

G. Tuyisenge (✉)
Department of Geography, Simon Fraser University, Burnaby, BC, Canada

P. T. Makanga (ed.), *Practicing Health Geography*, Global Perspectives on Health Geography, https://doi.org/10.1007/978-3-030-63471-1_8

example, highlights five dimensions that need to be considered: *availability* of services such as health facilities and the number of care providers; *accessibility*, which is related to the physical location where people live and seek health services and how they get to those services; the *accommodation* of clients' needs; *affordability* in terms of the ability to cover the direct and indirect costs of care; and *acceptability*, which is related to the personal characteristics that influence clients' attitudes towards health care (Penchansky and Thomas 1981; Evans et al. 2013).

In some cases, different terminologies are used interchangeably to reference Penchansky and Thomas' (1981) essential dimensions of access, which are also called the *5As* of access. For example, 'approachability' is sometimes used to refer to accessibility; 'adequacy' is used for availability, and 'accommodation' and 'appropriateness' are used interchangeably for acceptability (Gulliford et al. 2002; Levesque et al. 2013). Moreover, Saurman (2016) advocates for expanding the Penchansky and Thomas[2] concept in order to develop a better understanding of the complex nature of access. For example, some authors have advocated to add 'awareness' to the list of dimensions of access, to acknowledge its impact on access especially in the context of rural and remote settings with little exposure to health services and where populations might not be aware of health services available to them (Saurman 2016; Ward et al. 2015).

Access to Maternal Health care: The Context of Sub-Saharan Africa (SSA)

Maternal health remains one of the biggest challenges in the public health field. An estimated 830 women throughout the world die daily from pregnancy-related causes. The majority of these deaths are preventable (Alkema et al. 2016). It is reported that 99% of these deaths are experienced by women in developing countries, which speaks to the need for this population to receive improved and more timely access to skilled health care before, during, and after delivery (Filippi et al. 2006; Khan et al. 2006; Berry 2009). More specifically, Alkema et al. (2016) report that more than half of global maternal deaths occur in sub-Saharan Africa (SSA), a region that has disproportionately poor maternal outcomes due to the low socio-economic status of its population, poor health systems, poorly developed health infrastructures, low doctor-patient ratios, presence of health inequities, and high cost of health care. Tey and Lai (2013) argue that poor maternal outcomes in SSA are associated with poor access to maternal health services as well as poor utilization of those services. More than half of births in SSA are assisted by non-skilled birth attendants, such as Community Health Workers (CHWs), and take place outside health facilities. A lack of sufficient number of skilled pre-natal care and home-based deliveries are among the major causes of death during pregnancy and childbirth in most rural parts of SSA (AbouZahr 2014). Factors associated with distance to health facility, the cost of transport, and a woman and her family's (partner and extended family) perceived need to use health facilities have been reported to influence access to maternal health services.

There are many factors, other than those mentioned above, associated with poor access to maternal health services in SSA countries. A number of studies have been conducted on the barriers to and facilitators of access to maternal health services in SSA (Babalola and Fatusi 2009; Fotso et al. 2009; Tuyisenge et al. 2019). For example, Ricketts (2009) and Atuoye et al. (2015) have found that in Ghana, spatial access—which refers to the travel time and means to reach a health care facility—is affected by distance to the health facility, road conditions, the availability of transport and weather conditions. This is further emphasized by Arku et al. (2013), who found that poor spatial access to health facilities in Ghana makes it difficult for mothers to access to maternal health care in a timely fashion. Although spatial access to maternal health varies across regions in SSA, disparities are observed within and between many countries. Makanga et al. (2017) note that spatio-temporal factors should be considered when examining spatial access in the SSA context. They state that poor road conditions due to inclement weather have a significant impact on access to maternal health care, regardless of what mode of transport is used. They also note that most rural women in Mozambique walk or use public transport to reach to the nearest health facilities. These two modes of transport are not reliable during the wet weather season, at which time a drop in the numbers of people using health facilities has been observed.

Exploring Barriers and Facilitators to Community-Level Maternal Health Care Access in Rwanda

In Rwanda, primary health care services, including ANC and delivery services are provided at the health centres by nurses. In addition, each health centre coordinates the activities of CHWs operating in its catchment area (there are about 120 CHWs per health centre). The role of CHWs in improving access to maternal health care in the country, as a form of primary health care, has been undeniably remarkable and led to a considerable increase in the use of antenatal care in the past 10 years (Joharifard et al. 2012). The involvement of CHWs in maternal health contributed to an overall decline in MMR from 567 to 290 deaths per 100,000 live births between 2005 and 2015 (Bongaarts 2016). Rwanda is among the countries that have made tremendous progress towards the improvement of maternal health. The progress was observed in the early 2000s, as the country was rebuilding its health system and development infrastructure that had been destroyed during the 1994 Genocide. However, Rwanda's MMR remains among the highest worldwide, which calls for stronger efforts to be invested in the promotion of better maternal health outcomes.

In the remainder of this chapter, the author reports on the findings of interviews conducted between June and August, 2017 that explored Rwandan women's experiences of access to maternal health services offered by Assistante Maternelle de

Santé (ASMs–CHWs specifically responsible for maternal health). The author conducted these interviews through health centres in five districts: Nyarugenge and Gasabo districts in the capital city, Kigali; Gakenke and Rulindo districts in the Northern province; and Ruhango district in the Southern Province. Participants were recruited with the help of community health coordinators at the health centres in each district. In total, data was collected from ten health centres (two in each district): Bumbogo and Gatsata in Gasabo district, Gitega and Kimisagara in Nyarugenge district (Kigali); Ruli and Nemba in Gakenke; Shyorongi and Kinihira in Rulindo; and Byimana and Kinazi in Ruhango. Three of the ten health centres served an urban population—Gatsata, Gitega and Kimisagara—and the rest were located in rural parts of the country.

A total of 21 in-depth interviews were conducted with women between the ages of 19 and 39. The inclusion criteria were (1) women at least 18 years old, (2) be either pregnant or have had a healthy baby within 1 year, and (3) to have used village ASM services. Participants were both married and single, and their level of education ranged from the completion of primary education to a few years of secondary school. All but two had between one and five children between the ages of 6 months and 12 years. The two other participants were pregnant for the first time. Most reported farming as their main income-generating activity. The remaining few were stay home mothers, tradespeople, artisans or workers at wage-earning jobs. The following sub-section highlights factors that facilitate or impede the access to maternal health services at the community level as reported by the women. These factors are reported through the five dimensions of access introduced by Penchansky and Thomas (1981). Direct quotes (translated into English) from participants are also presented where possible (Table 8.1).

Availability of Services

Participants were asked about the availability of maternal health services in their communities. They stated that they have CHWs who are in charge of maternal health services known as ASMs who provided them with information and advice. Participants highlighted that they may report their pregnancy to the ASM, or an ASM may come to visit them to ensure they received antenatal care if pregnancy was suspected. Participants expressed that the ASMs in their communities familiarized them with the maternal health services available at the health centre and helped them to make appointments. According to one participant, an ASM is the first ‘to go’ person for maternal health issues.

Table 8.1 Participant voices from the interviews

Access dimension	Quotation from interview
Accommodation	*'In the morning I give a bath to my child and we go to see the ASM because if you were late, chances were that you miss her, they are busy with their own activities'.* *'After knowing that I was pregnant, the community health worker visited me. I asked her if I will be received at the health centre because I did not have a partner. She reassured me that I will be treated. I went, with a reference note from her and told my situation to the nurse and they received me. It was my first time to seek antenatal care during my first term of pregnancy'.*
Accessibility	*'She visited me when I was pregnant...she would remind me of the antenatal care appointments because she would get a reminder on her cell phone from the health centre. When my due date approached, she received a text message and came to remind me to be ready'.* *'It took me about three hours because of physical weakness and it is in mountains... I may be exhausted sometimes because when it was the market day; I went carrying some things on my head to sell...but nothing could stop me from coming'.*
Availability	*'If I had any problem, I went to the ASM to get advice from her. She showed me how I should handle things and I can't even come here at the health Centre without passing by her'.* *'For instance, you may be sick and if you call her, she would come to see you at your home with medications'.*
Affordability	*'Some pregnant women may deliver their babies without doing any antenatal care visit at the health centre. Sometimes it was because they did not have health insurance or did not have nice clothes to wear'.* *'Nurses criticized your hygiene in front of other mothers and they would laugh at you. It was embarrassing'. Another explained: 'My husband decided when to go... He told me to wait for a week to get paid so that he could be able to buy us new clothes before we came'.*
Acceptability	*'I changed my mind because there was a time when there were not at all people in charge of maternal health. A mother was on her own regarding her maternal health... now we have ASMs to help us... Your baby is given vaccination when you give birth at the health centre which you can't benefit from when you give birth at home'.* *'They are involved because they encourage you to do it as something important and they support you'.*

Accommodation

Participants shared their experiences regarding accommodation of their maternal health service needs by ASMs and health facilities. One participant explained how she organized her day when she needed to see an ASM. A few participants reported that even though they lived in the same communities as their ASMs, it was sometimes hard to get hold of them because they were usually involved in their own activities of daily life and did not have specific times when women could go to see them. This is particularly so in urban areas, where ASMs were more likely to be involved in paid labour outside the home. In these cases, women often had to call

ASMs on their cell phones to ask questions or schedule an appointment. This would present obvious challenges for those with no cell phones or those who could not afford the cost of making calls. Participants indicated that ASMs' roles included linking them to health facilities and supporting them to receive services that they may have been unfamiliar with.

In Rwanda, pregnant women are required to be accompanied by their spouses/ partners for their first visit to antenatal services in order to be tested for human immunodeficiency virus (HIV). Participants reported that this was a good strategy as they could learn their HIV status and make sure the baby was safe in case the mother was HIV+. However, they sometimes found it hard to get their partners to accompany them, particularly if the couple did not live together for reasons such as work. When participants were asked how well the health centres met their needs, most stated that the number of available nurses was limited. Thus, they had to spend many hours waiting for care, which discouraged their partners from accompanying them because it meant the partners would lose a full day of work. The support of ASMs in such cases was particularly welcome.

Accessibility

When asked about physical access to ASM services, participants shared that they could go to ASMs' homes when services were needed. Alternatively, the ASMs frequently visited their households to check on them when they were pregnant or check on their infants. Participants highlighted that having ASMs located close to them allowed them to get information about maternal health whenever they needed. As a result, they did not need to go health centres to address every single issue or concern, which was convenient given that most health centres were located far away. In addition, most participants said that they relied on ASMs to accompany them to health facilities, especially when they felt weak physically or were ready to deliver their babies. Participants also explained that access to clinics could be challenging during rainy weather '*because the roads got slippery to walk*', but this was an issue that ASMs were unable to rectify. In terms of physical accessibility to health centres, the time it took participants to travel to the nearest health facility ranged from 30 min to 3 h. When asked what mode of transportation they used, most reported that they walked. Another participant stated that some women were taken to health centres by traditional ambulance (a hand-woven stretcher) when they are unable to walk.

Affordability

Participants reported that the services provided by ASMs were free in communities. In health centres, antenatal care services were covered by health insurance. Participants had to pay for delivery and for the cost of an ambulance if they used it to be transferred from a health centre to a district hospital. However, participants shared different scenarios demonstrating that financial factors impacted the affordability of maternal health services. Participants highlighted the importance of having enough money to present themselves well at clinics.

Acceptability

Most participants acknowledged that having ASMs in their communities had given them a more positive attitude to maternal health services and the acceptability of accessing formal care services. ASMs' crucial roles included providing maternal health information about nutrition and physical activity, helping mothers to prepare for their babies, and encouraging pregnant women to attend antenatal care and to deliver in a facility. One participant explained that she had been reluctant to seek out antenatal care, but changed her mind after having five babies at home.

Other participants stressed the role that neighbours played in ensuring that they used maternal health services, thereby heightening acceptance of such care. Table 8.1 illustrates some of the participants' voices in regards to access to maternal health services offered by ASMs.

Discussion and Conclusion

The interview findings shared in the previous section show some of the ways through which the different dimensions of access are experienced in the context of maternal health. These dimensions are impacted by the characteristics of women plus the dynamics of places where they live (Kearns 1993), which in return impact their maternal outcomes. In other words, access to maternal care is shaped by different factors, including a combination of individual-level socio-economic and demographic factors of women's everyday lives and the larger socio-environmental factors that shape and create the places where they live. For example, a number of studies have reported on place-based disparities in access maternal health care within countries in the SSA region (Babalola and Fatusi 2009; McTavish et al. 2010). These disparities are mostly observed between rural and urban areas, due to the different socio-environmental factors that shape each of these settings, such as population density, number of health facilities and resources allocation (Schuurman 2009).

Disparities in access to maternal health care have been reported among women living in the same geographic area (Fotso et al. 2008). The findings of the interviews shared in this chapter show that in such cases these disparities are heavily driven by individual-level socio-economic factors that shape access. For example, a male partner's ability to purchase new clothing for mother and baby is a financial factor that shapes affordability. Similarly, the ability for this same partner to take time away from agricultural activity or paid labour to accompany a pregnant woman to an appointment at the health centre also shapes this particular dimension of access to maternal care. There is limited literature on male partner involvement in maternal health in SSA, and as a reported by Ditekemena et al. (2012), this involvement is impacted by different factors, including culture, the wait times, loss of income while in health centres and poor treatment of male partners by health professionals (see also Påfs et al. 2015). The interview findings shared here point to the importance of further exploring male partner involvement in shaping and determining women's access to maternal care. Such research should also explore the nature of ASMs' roles in overcoming this potential access barrier through undertaking measures such as accompanying women to health centre appointments or, as documented here, providing letters of introduction for un-accompanied women.

The perceived non-respectful treatment by health professionals as a barrier to care outside the community has been reported in our study, as well as its impact on the acceptability of maternal health services. Finlayson and Downe (2013) highlight that the utilization of antenatal care services is strongly associated with how women are treated by health professionals and respectful maternal health in general. It is thus not surprising that participants emphasized the importance of being able to see ASMs at the community-level who are familiar with the local culture, customs and norms around professional interaction. Similarly, Rosen et al. (2015) emphasize that when women are poorly treated by health professionals it decreases their trust in facility-based health services, which points even more to the importance of offering ASM care in the community. Consequently, this could translate into poor utilization of these services and overall poor acceptability of these services.

Although these interviews focused on ASM-based care in the community, larger community efforts to improve maternal health and ease access to care were revealed. For example, initiatives such as saving groups have been implemented in different communities to alleviate both affordability and physical accessibility barriers, as well as group purchases of stretchers. Reference to such initiatives clearly situated ASMs as members of local communities that also had resources that can contribute to facilitating access to maternal care. For the most part, participants praised ASMs for accommodating their needs, as they could reach out to them more easily without the need to go to the health centre for every inquiry. However, it is important to note that as volunteers and with the many tasks that they need to accomplish, ASMs are not always able to accommodate the maternal health needs of women. In this context Raven et al. (2015), call for more support and improved management of CHWs services by the health systems in SSA for better success of health services. The findings of the interviews reported on in this chapter confirm the importance of this call for support and improved management. Further to this, Singh and Sachs (2013) note that due to lack of funding and poor integration of CHWs in the health system in most SSA countries, CHWs get to operate in only a few regions of a country and in most cases their services are not formalized. Rwanda's ASM model provides an example of how to expand access into rural and remote areas and thus warrants further exploration in the context of strengthening the provision of community-based care in the SSA context.

References

AbouZahr, C. (2014). Progress and challenges in women's health: An analysis of levels and patterns of mortality and morbidity. *Contraception, 90*(6), S3–S13. https://doi.org/10.1016/j.contraception.2014.03.007.

Alkema, L., Chou, D., Hogan, D., Zhang, S., Moller, A. B., Gemmill, A., et al. (2016). Global, regional, and national levels and trends in maternal mortality between 1990 and 2015, with scenario-based projections to 2030: A systematic analysis by the UN Maternal Mortality Estimation Inter-Agency Group. *The Lancet, 387*(10017), 462–474.

Arku, G., Mkandawire, P., Luginaah, I., & Baiden, P. (2013). Healthcare access in three residential neighborhoods in Accra, Ghana. In *Spatial inequalities* (pp. 191–204). Dordrecht: Springer.

Atuoye, K. N., Dixon, J., Rishworth, A., Galaa, S. Z., Boamah, S. A., & Luginaah, I. (2015). Can she make it? Transportation barriers to accessing maternal and child health care services in rural Ghana. *BMC Health Services Research, 15*(1), 333.

Babalola, S., & Fatusi, A. (2009). Determinants of use of maternal health services in Nigeria-looking beyond individual and household factors. *BMC Pregnancy and Childbirth, 9*(1), 43.

Berry, N. S. (2009). Making pregnancy safer for women around the world: the example of safe motherhood and maternal death in Guatemala. See Hahn & Inhorn, 2009, 422–46.

Bongaarts, J. (2016). WHO, UNICEF, UNFPA, World Bank Group, and United Nations Population division trends in maternal mortality: 1990 to 2015. Geneva: World Health Organization, 2015. *Population and Development Review, 42*(4), 726.

Ditekemena, J., Koole, O., Engmann, C., Matendo, R., Tshefu, A., Ryder, R., & Colebunders, R. (2012). Determinants of male involvement in maternal and child health services in sub-Saharan Africa: A review. *Reproductive Health, 9*(1), 32.

Evans, D. B., Hsu, J., & Boerma, T. (2013). Universal health coverage and universal access. *Bulletin of the World Health Organization, 91*(8), 546–546A.

Filippi, V., Ronsmans, C., Campbell, O. M., Graham, W. J., Mills, A., Borghi, J., et al. (2006). Maternal health in poor countries: The broader context and a call for action. *The Lancet, 368*(9546), 1535–1541.

Finlayson, K., & Downe, S. (2013). Why do women not use antenatal services in low-and middle-income countries? A meta-synthesis of qualitative studies. *PLoS Medicine, 10*(1), e1001373.

Fotso, J. C., Ezeh, A., & Oronje, R. (2008). Provision and use of maternal health services among urban poor women in Kenya: What do we know and what can we do? *Journal of Urban Health, 85*(3), 428–442.

Fotso, J. C., Ezeh, A., Madise, N., Ziraba, A., & Ogollah, R. (2009). What does access to maternal care mean among the urban poor? Factors associated with use of appropriate maternal health services in the slum settlements of Nairobi, Kenya. *Maternal and Child Health Journal, 13*(1), 130–137.

Gulliford, M., Figueroa-Munoz, J., Morgan, M., Hughes, D., Gibson, B., Beech, R., & Hudson, M. (2002). What does 'access to health care' mean? *Journal of Health Services Research & Policy, 7*(3), 186–188.

Joharifard, S., Rulisa, S., Niyonkuru, F., Weinhold, A., Sayinzoga, F., Wilkinson, J., et al. (2012). Prevalence and predictors of giving birth in health facilities in Bugesera District, Rwanda. *BMC Public Health, 12*(1), 1049–1058.

Kearns, R. A. (1993). Place and health: towards a reformed medical geography. *The Professional Geographer, 45*(2), 139–147.

Khan, A. A., & Bhardwaj, S. M. (1994). Access to health care: A conceptual framework and its relevance to health care planning. *Evaluation and the Health Professions, 17*(1), 60–76.

Khan, K. S., Wojdyla, D., Say, L., Gülmezoglu, A. M., & Van Look, P. F. (2006). WHO analysis of causes of maternal death: a systematic review. *The Lancet, 367*(9516), 1066–1074.

Levesque, J. F., Harris, M. F., & Russell, G. (2013). Patient-centred access to health care: Conceptualising access at the interface of health systems and populations. *International Journal for Equity in Health, 12*(1), 18.

Makanga, P. T., Schuurman, N., Sacoor, C., Boene, H. E., Vilanculo, F., Vidler, M., et al. (2017). Seasonal variation in geographical access to maternal health services in regions of southern Mozambique. *International Journal of Health Geographics, 16*(1), 1.

McTavish, S., Moore, S., Harper, S., & Lynch, J. (2010). National female literacy, individual socio-economic status, and maternal health care use in sub-Saharan Africa. *Social Science & Medicine, 71*(11), 1958–1963.

Påfs, J., Musafili, A., Binder-Finnema, P., Klingberg-Allvin, M., Rulisa, S., & Essén, B. (2015). 'They would never receive you without a husband': Paradoxical barriers to antenatal care scale-up in Rwanda. *Midwifery, 31*(12), 1149–1156.

Penchansky, R., & Thomas, J. W. (1981). The concept of access: Definition and relationship to consumer satisfaction. *Medical Care, 19*, 127–140.

Raven, J., Akweongo, P., Baba, A., Baine, S. O., Sall, M. G., Buzuzi, S., & Martineau, T. (2015). Using a human resource management approach to support community health workers: Experiences from five African countries. *Human Resources for Health, 13*(1), 45.

Ricketts, T. C. (2009). Accessing health care. A Companion to health and medical geography, 521–539.

Rosen, H. E., Lynam, P. F., Carr, C., Reis, V., Ricca, J., Bazant, E. S., & Bartlett, L. A. (2015). Direct observation of respectful maternity care in five countries: A cross-sectional study of health facilities in east and southern Africa. *BMC Pregnancy and Childbirth, 15*(1), 306.

Saurman, E. (2016). Improving access: Modifying Penchansky and Thomas's theory of access. *Journal of Health Services Research & Policy, 21*(1), 36–39.

Schuurman, N. (2009). Chapter 4: The effects of population density, physical distance and socio-economic vulnerability on access to primary health care in rural and remote British Columbia, Canada. In *Primary health care: People, practice, place* (pp. 57–74). Farnham: Ashgate.

Singh, P., & Sachs, J. D. (2013). 1 million community health workers in sub-Saharan Africa by 2015. *The Lancet, 382*(9889), 363–365.

Tey, N. P., & Lai, S. L. (2013). Correlates of and barriers to the utilization of health services for delivery in South Asia and sub-Saharan Africa. *The Scientific World Journal, 2013*, 1.

Tuyisenge, G., Hategeka, C., Kasine, Y., Luginaah, I., Cechetto, D., & Rulisa, S. (2019). Mothers' perceptions and experiences of using maternal health care services in Rwanda. *Women & Health, 59*, 68–84.

Ward, B., Humphreys, J., McGrail, M., Wakerman, J., & Chisholm, M. (2015). Which dimensions of access are most important when rural residents decide to visit a general practitioner for non-emergency care? *Australian Health Review, 39*(2), 121–126.

Chapter 9
Using Participatory Photovoice Research to Understand Food Allergy Risk in Accra, Ghana

George A. Atiim, Elijah Bisung, and Susan J. Elliott

Abstract Chronic noncommunicable diseases (NCD) are becoming a significant public health issue in Ghana. While a national policy response exists to prevent the burden of NCDs, little attention is given to food allergy, a growing allergic disease in much of the world. Understanding the illness experience of individuals can provide useful information on place-based factors that facilitate or militate against health and well-being, as well as reveal the agency of individuals to modify their social context in-spite of illness. This chapter highlights the application of photovoice, a participatory qualitative research approach to understand the experiences of people with food allergy in an urban setting in a low- and middle-income country context (Ghana). We identified the multiple spaces of exposure to children, as well as the institutional and policy barriers that pose a difficult challenge for the diagnosis, treatment and management of food allergy in a developing country setting. Despite challenges, photovoice provides an opportunity for linking food allergy experiences with the social circumstances within which people live and work.

Keywords Photovoice · Food allergies · Qualitative research · Social context

G. A. Atiim (✉)
United Nations University International Institute for Global Health (UNU-IIGH), Kuala Lumpur, Malaysia

E. Bisung
School of Kinesiology and Health Studies, Queen's University, Kingston, ON, Canada

S. J. Elliott
Department of Geography and Environmental Management, University of Waterloo, Waterloo, ON, Canada

P. T. Makanga (ed.), *Practicing Health Geography*, Global Perspectives on Health Geography, https://doi.org/10.1007/978-3-030-63471-1_9

Background

Chronic noncommunicable diseases (NCD) are becoming a significant public health issue in Ghana and African countries, adding to an already existing burden from communicable diseases. In Ghana, health records show a high number of reported cases of asthma, cardiac disease, hypertension, diabetes mellitus and sickle cell at outpatient health facilities between 2011 and 2014 (Ghana Health Service 2015). While a national policy response exists to prevent, and mitigate the burden of NCDs, little attention is on food allergy, a growing allergic disease that affects 7–10% of people in the world, particularly in western societies (Soller et al. 2015; Osborne et al. 2011).

Although national data on food allergy prevalence in Ghana is lacking, a growing body of research (Atiim et al. 2018; Afaa et al. 2017; Obeng et al. 2011) suggests the need to give some priority to food allergies. For example, we know that 11% of schoolchildren self-report a food allergy (Obeng et al. 2011), with studies in Ghana (Atiim et al. 2018), and elsewhere (Cummings et al. 2010; Ostblom et al. 2008) documenting the psychosocial, and economic impact on those with allergies, and/or their caregivers as they negotiate daily living. In the most extreme cases, food allergy can induce anaphylaxis, a serious life-threatening condition that can lead to death when left untreated (Hochstadter et al. 2016).

In the absence of a cure, prevention remains a key strategy with complete avoidance of allergens, and the use of an epinephrine auto injector (EAI)[1] when anaphylaxis do occur, considered the critical steps towards preventing and managing food-induced allergic reactions (Boyce et al. 2010). However, non-compliance can be an issue; with studies reporting that a strict avoidance of allergens, carrying or even use of epinephrine—the lifesaving medicine—is often suboptimal (Waserman et al. 2017; Sheikh et al. 2015; Greenhawt et al. 2009). For example, among college students, only 33% report avoiding allergens at all times, with only 10% carrying epinephrine (Greenhawt et al. 2009). Another study found that, even if persons at risk of anaphylaxis had an EAI, many did not use them when their reactions reoccurred (Sheikh et al. 2015).

As with many chronic health issues, the burden of disease and the illness experience often increases over time. Understanding the illness experience of individuals can provide useful information on place-based factors that facilitate or militate against health and well-being, as well as reveal the agency of individuals to modify their social context in-spite of illness. In this research, we use photovoice, a participatory qualitative research approach to understand the experience of people with food allergy through photographs in cosmopolitan Accra, Ghana. In capturing photos, coupled with the narratives and accompanying interpretations from the perspective of affected individuals, we obtain insights of the sociocultural conditions shaping the allergy experience. This chapter focuses on the utility and challenges of using photovoice to study food allergies in Ghana.

[1] A hand-held medical device for administering a measured dose of epinephrine through injecting.

Research Context

This study was undertaken in Accra in the Greater Accra Region (GAR), one of the most economic and cosmopolitan areas in Ghana. Our choice of the GAR was informed in part by the knowledge that allergies appear to be higher in urban areas and follows along a socioeconomic gradient (Viinanen et al. 2005; Hijazi et al. 2000). As such, we expected food allergies to be more common in Accra, considering that it is the most urbanized in Ghana (Ghana Statistical Service 2012) with the GAR having more than half the population falling within the highest wealth quintile, compared to just 2% in the Northern and Upper East regions (Ghana Statistical Service (GSS) et al. 2015).

Like many other NCDs, food allergies are partly driven by economic transformation and the globalization process. The GAR is a melting pot for much of these ongoing transformations, leading to lifestyle and behavioural changes. Recent data report high rates of being overweight and obese (55.2%) in Greater Accra compared to a national average of 43% in adults (Ofori-Asenso et al. 2016). It is in this context of ongoing socioeconomic inequalities and changing disease landscape that the GAR provides an ideal setting to explore this emerging health risk and the factors that influence risk perceptions around food allergy.

Photovoice Research Method

Photovoice is a growing action-research method that promotes participation and social change by allowing individuals and/or communities to identify, and analyse their experiences or community problems through photography (Bisung et al. 2015). From both a theory and practice lens, photovoice is about giving voice to participants, the active co-creation of knowledge, and empowerment to transform local problems (Masterson et al. 2018; Castleden et al. 2008; Wang and Burris 1997).

As originally espoused by Wang and Burris (1997) and reiterated by others (Budig et al. 2018; Kingery et al. 2016; Fournier et al. 2014), photovoice seeks to fulfil three critical goals namely: (1) to enable people to document, reflect and discuss important issues at a personal or community level; (2) promote dialogue and share knowledge around an issue and (3) reach decision makers for public policy change. Photographs have the ability to navigate both language and literacy barriers (Kingery et al. 2016), as well as reveal critical information about conditions within communities that shape individual outcomes (Belon et al. 2016). Thus, photovoice has the potential to reveal how people and their contexts may interact to influence health outcomes.

Considering that food allergy has both direct and indirect impacts on psychosocial and economic well-being (Cummings et al. 2010; Ostblom et al. 2008), photovoice could be a useful method for exploring personal and community-level factors that shape the experience of living with food allergies, particularly in contexts where they maybe unrecognized as an important health challenge (Atiim et al. 2017). However, while photovoice has been used to explore risk perception, disparities across management and public health discourses on asthma (Yarbrough et al. 2016; Evans-Agnew 2016, 2018), few applications exist for food allergy (see Valentine and Knibb 2011 for exception). This is particularly the case in contexts such as sub-Saharan Africa, although there is an abundance of it utility to examine general health needs (Kingery et al. 2016), water and sanitation challenges (Bisung and Elliott 2019), children and HIV (Fournier et al. 2014) and human well-being in socioecological systems (Masterson et al. 2018). As prevention remains critical to avoiding an allergic reaction, photovoice provides an opportunity for allergic individuals and those indirectly affected (e.g. parents/guardians) to co-lead research into their lived experience, while identifying the drivers that facilitate or act as a barrier to shape health experiences, practices and behaviours.

Method: Putting Photovoice into Practice

Between June and August 2015, eight (8) parents/guardians of children with a food allergy, mostly females ($n = 6$) and between the ages of 25 and 40 years were recruited to implement a photovoice project in the Accra metropolis of the GAR, Ghana. Participants were recruited based on either (a) a self-report of their child having a food allergy or (b) a self-report of physician diagnosis or suspected allergy and (c) a report of at least two relevant symptoms (d) occurring within 2 h of contact with the allergen (Boyce et al. 2010). The sample size is consistent with previous research (Bisung et al. 2015), appropriate and adequate to generate in-depth information (photographs and narratives) on the subject matter in-focus—i.e. preventing allergic reactions (Miles and Huberman 1994; Curtis et al. 2000). Further, the sample size is ideal for keeping and managing selected photographs in face-to-face-interviews and/or group discussions (Dennis et al. 2009; Curtis et al. 2000).

Participants were handed disposable cameras (with a single use film) for a week and tasked to capture and record through photographs their perceptions and experiences managing food allergy in the community (e.g. 'when you think about your children food allergy, what comes to your mind'?). Considering the duration of photo taking exercise, daily reminders through text messages were sent to participants to encourage and remind them to take photographs. Prior to photo taking, consent was obtained, and each individual was trained on how to use the camera and instructed on privacy and ethical issues that may likely arise from taking photographs. While not prescriptive, we encouraged participants to reflect on practices, behaviours and interactions in a variety of settings in order to critically capture photographs that appropriately illustrated their lived experiences.

In total, between 10 and 18 photographs were captured by each participant which were subsequently developed and printed. Participants received copies of the photographs and asked to choose four photos that appropriately resonate with their experiences managing food allergy. Subsequently, a semi-structured in-depth interview, conducted simultaneously in English and Twi (one of the local languages widely spoken) was held to discuss their practices, behaviours and interactions regarding management through the lens of the selected photographs. Interviews lasted between 30 and 45mins focusing around their motivation, reasoning and meanings associated with each photograph. Questions that guided the interviews included: (1) what is in this picture and where was it taken? (2) why is photo important to understanding food allergy? (3) how does it relate to well-being in the community? and (4) what could be done to address the issues or challenges outlined in the photos? All conversations were audio-taped with prior permission from participants.

Analysis

Next, all interviews were transcribed verbatim. Subsequently, the photos and transcripts were imported into NVivo11, a qualitative data analysis and management software package. Participants identified themes for photographs were initially used for coding, with additional themes included from the transcripts as coding progressed. Where a photograph captures more than a theme, it was placed in more than one category. To ensure credibility and consistency of data (Miles and Huberman 1994), coding was undertaken with the assistance of a second coder. In cases of differential coding, discrepancies were discussed and agreement reached by consensus. Additionally, field notes documenting observations including casual remarks during face-to-face interviews were also analysed and used to complement the identified themes.

Results

The aspiration of a photovoice project is to create opportunities for people to document, reflect and discuss critical concerns facing them in the context of their community. In doing so, photovoice promotes dialogue and knowledge sharing with the aim of reaching policy makers and change policy. While the overall project revealed several important insights around food allergens (including novel allergens) and lived experiences as it relates to diagnosis and management of allergy, we focus on four lessons drawn from this participatory approach through the lens of our participants.

Promoting Knowledge, Attitudes and Household Practice Changes

Despite a context of inadequate healthcare infrastructure for treatment and management of food allergies in Ghana (Atiim et al. 2018), the photos and narrative data described the ways through which participants responded to the knowledge that their children may not be safe and exposure could occur anywhere as illustrated below:

> This activity gets you to really think about what you're doing. So for me, it was the fact, they are not safe. Reaction can strike anywhere which was a sobering thought for me throughout this process—Silvia, 35 years, mother of 4 year allergic girl

As a result several photographs highlight daily practices and activities implemented to prevent allergic reactions including (but not limited to) reading labels, keeping utensils clean, separate eating utensils; keeping medicines accessible and maintaining strict allergen avoidance (see Plate 9.1). These actions to prevent exposure vary across multiple contexts—the home, school, social gatherings—and partly demonstrate the agency and capacities of participants in illness management.

"Always looking at labels to identify if there is an allergen.This one for example has soya in it.

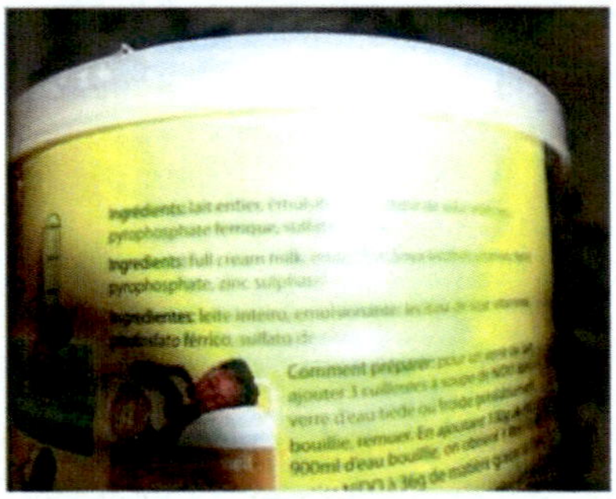

"I am cleaning every utensil before and after use to avoid mixing, and contamination"

"He has his own eating stuffs to prevent any mishaps".

"Peanuts are foods we completely avoid now"

Plate 9.1 Daily practices and activities to manage food allergies

Through systematically identifying and documenting potential risks, and several accompanying coping actions, participants emphasized how they have now become increasingly conscious of the sustained measures instituted to forestall future exposures. Prior to photovoice, none indicated deliberately recording and re-evaluating the effectiveness of mechanisms put in place to cope with and prevent food allergic reactions. As an outcome of the process, many observed an improvement in allergy management, with most reporting a noticeably decline in children symptoms of allergies. While also acknowledging some challenges (e.g. water shortages, non-compliance), many participants mentioned that coping strategies have generally been effective.

Advocates for Social Change on the Visibility of Food Allergy

Participants reported that convincing community members of the validity of their children allergy can be a challenge, a theme consistent within the broader research findings (Atiim et al. 2018). Some saw the photovoice activity as a powerful tool for some miniature validation of children allergy and an opportunity to educate others. The photo taking activity provided strong leverage for some participants to engage and to communicate with curious community members about the objective of this project, but more importantly, how it relates to their child allergy. For example, one father of a 3-year allergic girl describes an instantaneous teachable moment noting '*people will often asked why are you doing this? It was a chance to educate them not just about what you're doing but also that my girl's condition is so true or important that they are doing research about it right here*'.

Similar sentiments were expressed (see Plate 9.2), with some participants intimating the cameras created opportunities to address people curiosities about the photo taking and a space for conversations around critical needs, experiences and challenges of the allergic population and their families. In doing so, photovoice helped to extend participants responsibility as social actors within their community, even if this change was starting at an individual level in their neighourhoods or at the workplace.

> The inscription captured me. What have you done today to make you feel proud it said? And for me, it gave the chance to speak about my child food allergy, which on an ordinary day, I would not rise up and start telling people my son has an allergy. People were I think, genuinely curious and that lead to conversations which wouldn't happen on a typical day—Mary, 36 years, mother of 10 year old boy with peanut allergy

Plate 9.2 A photo highlighting advocacy activities of participants

Enhanced Understanding of Risk and Spaces of Exposure

At the start of the project, while all participants recognized offending foods (see Plate 9.3) as the immediate hazards children faced, photovoice helped to identify the influence of external environments[2] on allergic reactions reoccurring. For many, risk is ever-present although with varying degrees across various spaces of interactions, with the most frequently captured space of exposure in descending order being the school ($n = 4$), restaurants ($n = 2$), place of worship, and home environment (see Plate 9.4). While allergen avoidance was the most pervasive prevention strategy, the self-realization that exposure can occur across this spectrum of living environments led to a re-evaluation of prevention strategies to mitigate risk of an allergic reaction occurring as illustrated by the quote below:

> I knew it could happen to him anywhere but never really thought 'paaa' [deeply] about them having the same effect. It's making me to rethink about how we are treating him. Our efforts is only keeping him safe at home and school but he should be safe wherever he is—Rosalyn, 31 years, mother of 7 year old with egg allergy

Hitherto, much preventive action—i.e. avoidance of allergen, reading labels, informing others of child's allergy—existed predominantly at the home and community level, with little emphasis on equally critical spaces of interactions with potential to trigger reactions. Following photovoice, there was a sense of urgency for measures that covers 'all possible loopholes' that may put children in danger of a reaction, with most intending to begin processes of engaging other key personnel in

[2] We conceptualized this as all other settings or spaces outside the confines of participants' households.

Plate 9.3 Selected examples of photos illustrating offending food allergens: (**a**) a crate of eggs, (**b**) a plate containing soya, (**c**) peanut at a local market and (**d**) fish

educational (e.g. class teacher, school health staff, head teacher), religious (e.g. teachers, religious leaders), and eatery (e.g. waiters, chefs) settings about their children's allergies. For example, one parent commenting on this said, '*the eye opener for me is, I have to take the same seriousness at home, and tell his school and our worship leaders about what he can eat and what they cannot give him*'. With similar sentiments, participants understood better that preventing future allergic reactions required moving beyond the safety of the home.

Strengthening Health Institutions and Personnel for Allergy Management

Photographs depicting a range of health services (e.g. hospitals and clinics [*n* = 6], and pharmacies) both public and private drew attention to an important need, the strengthening of health system infrastructure including but not limited to

Plate 9.4 Perceived spaces that likely poses risk to the allergic population

knowledge and diagnostic competencies, diagnostic tools, and greater attention to 'unpopular health problems'. While the photos (see Plate 9.5a,b) chosen initially were to illustrate the 'where' of health service availability and accessibility, constrains within the healthcare and policy environment immediately dominated discussions around the meanings associated with these photographs. For example, when discussing Plate 9.5a, a participant noted: '*it reminded me of our going and coming, and still we didn't know what the cause was, and also how we don't really have the right personnel. Lots of them did not really know what to do. That's was our period of worry*'. This sentiment underscores how the lack of qualified allergists in the local healthcare system may have contributed to prolonged delays in diagnosing her daughter's allergy, with its accompanying experiences of frustrations and anxiety.

Equally important issue was descriptions of food allergies as 'unpopular health problems', in an apparent reference to the 'seeming' lack of attention in the healthcare system. For example, one participant intimated: '*I was thinking about people with power. It looks like they aren't interested in food allergy because it's not popular*'. By capturing and highlighting the policy context, attention is drawn to how health governance structures and practices produce and sustain priorities on particular health issues while simultaneously neglecting others—in this case, food allergies as a public health concern. Additionally, dissatisfaction with healthcare workers

"Our hospitals don't have the right tools, and personnel. So it's a little difficult especially the process of finding out the problem in the first place"

"It isnt really about the people here. If those in authority don't resource them, or care about these conditions, we cant expect them to perform magic there".

"When things really didn't go well at the beginning at the clinics, the pharmacy was where we turned. You described the symptoms and they provided you with some medicine for relief. You took whatever they gave because at this stage, all you wanted is some reliefs for you child".

Plate 9.5 (a) (b) Health system drivers of diagnosis, treatment and management of food allergy

prognosis resulted in an increased self-management and reliance on pharmacies, both licensed and unlicensed for children food allergies (see Plate 9.5b). While this did not necessarily improve children health, nonetheless, it presented an alternative space to explore potential solutions for children allergy.

Challenges Putting Photovoice into Practice

Photovoice as a participatory research has challenges in its implementation. As observed by others, it is a time intensive process, both for participants and researchers (Budig et al. 2018; Bisung et al. 2015). In this research, participation required spending time in photography and ethics training prior to the actual photo taking exercise. Post-photo taking, it also required reasonable time for interviewing. As such, the time intensive commitment can be a constrain, especially for participants who must find ways to situate this exercise in the context of competing daily interests at home, work or even within the community. A great deal of commitment may therefore be required from both participants and researchers in order to successfully implement a photovoice project (Bisung et al. 2015).

Further, both researchers and participants are sometimes confronted with a dilemma of choosing from the collection of printed photographs, a specific number (in this case of this research, 4 each) of 'best' photos for face-to-face interviews or group discussion. Photovoice projects have the tendency to sometimes produce an abundance of compelling pictures which may not be identified by participants as entirely relevant for discussion. While it is understandable to restrict the number of photos to ensure interviewing or discussions are manageable, by doing so, we could miss opportunities to learn from these unused photographs—that is the 'unselected' is denied attention when designing and implementing larger interventions (Bisung et al. 2015).

In addition, ethical concerns are increasingly at the forefront of photovoice projects as scholars grapple with questions around what can and cannot be photographed. For example, while an ethics training session on photography was held for all participants, in a few instances, the research team with guidance from local gatekeepers made decisions to exclude some selected photos by participants on grounds of either a lack of consent before the photo was taken or the anonymity of person in the photograph could not be guaranteed. This was not a decision affected participant often agreed with researchers but following some considerable discussions, replacement photos were often chosen.

Conclusion

Research on food allergy is nascent though marginal compared to the body of work on NCD in Africa. While studies are beginning to record prevalence (see Basera et al. 2015; Gray et al. 2014; Obeng et al. 2011), the social context of food allergy is characteristically missing in this body of research. Using a photovoice participatory approach, we have a better understanding of the active role participants play through initiating actions to safeguard their children from experiencing an allergic reaction. Through this approach, we now understand the multiple spaces of risk exposure to children, with the understanding that risk management for those living with a food allergy ought to extend beyond the confines of the home environment. Additionally, attention is drawn to the 'hidden barriers' operating both at institutional (e.g. healthcare) and policy levels that poses a difficult challenge for the diagnosis, treatment and management of food allergy in a developing country setting.

This photovoice project in Accra allowed participants to identify, record and articulate the central issues affecting the experience of food allergy. While the research did not deliberately set out to use photovoice to reach policy makers about the concerns of participants, nonetheless, decision makers stand to benefit immensely from listening to the voices of participants in order to guide the transformation of health systems that appropriately reflects and addresses the needs of those in need of critical health services.

Acknowledgement *Disclaimer*: The views expressed herein are those of the author(s) and do not necessarily reflect the views of the United Nations University.

References

Afaa, T. J., Afrane, A. K. A., & Etwire, V. (2017). Gastrointestinal food allergy in Ghanaian children: A case series. *Ghana Medical Journal, 51*(3), 138–142.

Atiim, G. A., Elliott, S. J., & Clarke, A. E. (2017). If we are waiting for the numbers alone, we miss the point: A qualitative study of the perceived rise of food allergy and associated risk factors in the Greater Accra Region, Ghana. *Global Health Research and Policy, 2*, 20.

Atiim, G. A., Elliott, S. J., & Clarke, A. E. (2018). "Ne nnipaduammpe" (the body hates it): Exploring the lived experience of food allergy in sub-Saharan Africa. *Social Science & Medicine, 205*, 72–81.

Basera, W., Botha, M., Gray, C. L., Lunjani, N., et al. (2015). The South African food sensitisation and food allergy population-based study of IgE-mediated food allergy: Validity, safety and acceptability. *Annals of Allergy, Asthma & Immunology, 115*, 113–119.

Belon, A. P., Nieuwendyk, L. M., Vallianatos, H., & Nykiforuk, C. I. J. (2016). Perceived community environmental influences on eating behaviours: A photovoice analysis. *Social Science & Medicine, 171*, 18–29.

Bisung, E., & Elliott, S. J. (2019). Community water improvement and health: A pre-post photovoice intervention study in Kenya. *Habitat International, 85*, 14–20.

Bisung, E., Elliott, S. J., Abudho, B., Abudho, B., Karanja, D. M., & Schuster-Wallace, C. J. (2015). Using photovoice as a community based participatory research tool for changing water, sanitation and hygiene behaviours in Usoma, Kenya. *BioMed Research International, 2015*, 903025, 10p.

Boyce, J. A., Assa'ad, A., Burks, W. A., Jones, S. M., Sampson, H. A., Wood, R. A., et al. (2010). Guidelines for the diagnosis and management of food allergy in the United States: Report of the NIAID-sponsored expert panel. *The Journal of Allergy and Clinical Immunology, 126*(6), S1–S43.

Budig, K., Diez, J., Conde, P., Sastre, M., Hernan, M., & Franco, M. (2018). Photovoice and empowerment: Evaluating the transformative potential of a participatory action research project. *BMC Public Health, 18*, 432.

Castleden, H., Garvin, T., & Huu-ay-aht First Nation. (2008). Modifying photovoice for community based participatory indigenous research. *Social Science & Medicine, 66*, 1393–1405.

Cummings, A. J., Knibb, R. C., King, R. M., Lucas, J. S., Lucas, J., & Wellcome, S. H. (2010). The psychosocial impact of food allergy and food hypersensitivity in children, adolescents and their families: A review. *Allergy, 65*(8), 933–945.

Curtis, S., Gesler, W., Smith, G., & Washburn, S. (2000). Approaches to sampling and case selection in qualitative research: Examples in the geography of health. *Social Science & Medicine, 50*(2), 1001–1014.

Dennis, S. F., Gaulocher, S., Carpiano, R. M., & Brown, D. (2009). Participatory photo mapping (PPM): Exploring an integrated method for health and place research with young people. *Health & Place, 15*, 466–473.

Evans-Agnew, R. A. (2016). Asthma management disparities: A photovoice investigation with African American youth. *The Journal of School Nursing, 32*(2), 99–111.

Evans-Agnew, R. A. (2018). Asthma disparity photovoice: The discourses of black adolescent and public health policymakers. *Health Promotion Practice, 19*(2), 213–221.

Fournier, B., Bridge, A., Kennedy, A. P., Alibhai, A., & Konde-Lule, J. (2014). Hear our voices: A photovoice project with children who are orphaned and living with HIV in a Ugandan group home. *Children and Youth Services Review, 45*, 55–63.

Ghana Health Service. (2015). *2014 Annual health report*. Accra: Ghana Health Service.

Ghana Statistical Service (GSS). (2012). *2010 Population & housing census summary report of final results*. Accra: Ghana Statistical Service.

Ghana Statistical Service (GSS), Ghana Health Service (GHS), ICF International. (2015). *Ghana demographic and health survey* (p. 2014). Rockville, MD: GSS, GHS, and ICF International.

Gray, C. L., Levin, M. E., Zar, H., Potter, P. C., et al. (2014). Food sensitization and allergy in South African children with atopic dermatitis. *Paediatric Allergy and Immunology, 25*, 572–579.

Greenhawt, M. J., Singer, A. M., & Baptist, A. P. (2009). Food allergy and food allergy attitudes among college students. *The Journal of Allergy and Clinical Immunology, 124*(2), 323–327.

Hijazi, N., Abalkhail, B., & Seaton, A. (2000). Diet and childhood asthma in a society in transition: A study in urban and rural Saudi Arabia. *Thorax, 55*, 775–779.

Hochstadter, E., Clarke, A., De Schryver, S., LaVieille, S., Alizadehfar, R., Joseph, L., et al. (2016). Increasing visits for anaphylaxis and the benefits of early epinephrine administration: A 4-year study at a pediatric emergency department in Montreal, Canada. *The Journal of Allergy and Clinical Immunology, 137*(6), 1888–1890.

Kingery, F. P., Naanyu, V., Allen, W., & Patel, P. (2016). Photovoice in Kenya: Using a community-based participatory research method to identy health needs. *Qualitative Health Research, 26*(1), 92–104.

Masterson, V. A., Mahajan, S. L., & Teno, M. (2018). Photovoice for mobilizing insights on human well-being in complex social-ecological systems: Case studies from Kenya and South Africa. *Ecology and Society, 23*(3), 13. https://doi.org/10.5751/ES-10259-230313.

Miles, M. B., & Huberman, A. M. (1994). *Qualitative data analysis: An extended sourcebook* (2nd ed.). Thousand Oaks, CA: Sage.

Obeng, B. B., Amoah, A. S., Larbi, I. A., Yazdanbakhsh, M., van Ree, R., Boakye, D. A., & Hartgers, F. C. (2011). Food allergy in Ghanaian schoolchildren: Data on sensitization and reported food allergy. *International Archives of Allergy and Immunology, 155*(1), 63–73.

Ofori-Asenso, R., Agyeman, A. A., Laar, A., & Boateng, D. (2016). Overweight and obesity epidemic in Ghana—A systematic review and meta-analysis. *BMC Public Health, 16*, 1239.

Osborne, N. J., Koplin, J. J., Martin, P. E., Gurrin, L. C., Lowe, A. J., Matheson, M. C., et al. (2011). Prevalence of challenge-proven IgE mediated food allergy using population-based sampling and predetermined challenge criteria in infants. *The Journal of Allergy and Clinical Immunology, 127*(3), 668–676.

Ostblom, E., Egmar, A. C., Gardulf, A., Lilja, G., & Wickman, M. (2008). The impact of food hypersensitivity reported in 9-year-old children by their parents on health related quality of life. *Allergy, 63*(2), 211–218.

Sheikh, A., Dhami, S., Regent, L., Austin, M., & Sheikh, A. (2015). Anaphylaxis in the community: A questionnaire survey of members of the UK anaphylaxis campaign. *JRSM Open, 6*, 1–6.

Soller, L., Ben-Shoshan, M., Harrington, D. W., Knoll, M., Fragapane, J., Joseph, L., et al. (2015). Adjusting for non-response bias corrects overestimates of food allergy prevalence. *The Journal of Allergy and Clinical Immunology, 3*(2), 291–293, e2.

Valentine, A. Z., & Knibb, R. C. (2011). Exploring quality of life in families of children living with and without a severe food allergy. *Appetite, 57*(2), 467–474.

Viinanen, A., Munhbayarlah, S., Zevgee, T., Narantsetseg, L., Naidansuren, T., Koskenvuo, M., & Terho, E. O. (2005). Prevalence of asthma, allergic rhinoconjunctivitis and allergic sensitization in Mongolia. *Allergy, 60*(11), 1370–1377.

Wang, C., & Burris, M. A. (1997). Photovoice: Concept, methodology, and use for participatory needs assessment. *Health Education & Behavior, 24*, 369–387.

Waserman, S., Avilla, E., Ben-Shoshan, M., Rosenfield, L., et al. (2017). Epinephrine autoinjectors: New data, new problems. *The Journal of Allergy and Clinical Immunology: In Practice, 5*(5), 1180–1190.

Yarbrough, M., Blumenstock, J., Warren, C., Dyer, A., et al. (2016). SMART (Student Media-based Asthma Research Team): Engaging adolescents to understand asthma in their communities. *Progress in Community Health Partnerships: Research, Education, and Action, 10*(4), 523–532.

Chapter 10
Determinants of Maternal Health in Regions of Southern Mozambique

Tabassum Firoz, Jerry Spiegel, and Prestige Tatenda Makanga

Abstract This chapter focuses on the circumstances and the wider set of forces and systems shaping the conditions of daily life, a concept increasingly summarized as the determinants of health. We begin by describing how these determinants encompass political, economic, social, cultural and environmental dimensions and the need to examine the pattern of interactions at different scales, and how thinking about developing this focus has evolved. We present critical perspectives in understanding the determinants and determination of health. Using the example of Mapping Outcomes for Mothers (MOM), a study undertaken in Mozambique, we illustrate why the historical narratives of these determinants are imperative in the analysis of health. The chapter ends with a discussion of how the inclusion of a consideration of determinants can accelerate progress through multi-sectoral collaboration and delivery of care that is not limited to the health sector.

Keywords Health determinants · Determinations of health · Maternal health · Geographical determinants of health

Introduction

Historically, health outcomes have been predominantly viewed within the biomedical paradigm. Such focus on health at the scale of the human body, organ systems and subunits have reinforced framing that are fundamentally reliant on biomedical perspectives to explain health outcomes. However, with increasing awareness of

T. Firoz (✉)
Department of Medicine, Yale School of Medicine, Bridgeport Hospital,
Yale New Haven Health, Bridgeport, CT, USA

J. Spiegel
School of Population and Public Health, University of British Columbia,
Vancouver, BC, Canada

P. T. Makanga
Place Alert Labs (PALs), Department of Surveying and Geomatics,
Midlands State University, Gweru, Zimbabwe

P. T. Makanga (ed.), *Practicing Health Geography*, Global Perspectives on Health Geography, https://doi.org/10.1007/978-3-030-63471-1_10

markedly different distributions of disease among and within populations, it is now recognized that health outcomes are not confined to only clinical, biological, or individual risk factors (March and Susser 2006). Rather, health and well-being are best conceived as an 'eco-bio-social' phenomena situated within broader contexts of society, culture, ecology, and political systems (Mansfield 2008). The local contexts of places and peoples' everyday lives are, therefore, crucial when considering health outcomes and health policy development.

Place and geographic context have been long recognized as important influences on health (Dummer 2008). The notion of regions of health is well established in studies on the geography of health and has included wider interpretation of health phenomenon in identifying the nature of different places (Collins 2002). Geography has been used to study many aspects of health such as health service utilization, disease surveillance including determinants of health and disease, and health inequalities.

In recent years, there has been a shift away from the biomedical model in geography and towards a holistic perspective that studies the relationship between health, context of places and the connectedness between places. Approaches have evolved to incorporate the application of social theory, and critical perspectives on the geographies of health (Jackson and Neely 2014).

Current developments in health geography include a stronger focus on health inequalities and the impact of spatial and social marginalization on health (Asthana et al. 2002). Landscape, therefore, can serve as a metaphor for 'the complex layering of history, social structure and built environment that converge in particular places' (Hoey 2007) and be utilized as a way of interrogating the relationships between social actors across multiple spatial and temporal scales (King 2009).

Social Determination of Health

In the era of the Sustainable Development Goals, there has been increased recognition that improving health and achieving health equity requires examination of the social, cultural, political, economic, and environmental factors that influence health. Awareness of extensive disparities regarding how such factors are distributed globally, among and within nations, was heightened a decade earlier through the work of the Commission on the Social Determinants of Health established by the World Health Organization [WHO] (Marmot et al. 2008).

The WHO defines the social determinants of health as the conditions in which people are born, grow, work, live, and age, and the wider set of forces and systems shaping the conditions of daily life (Marmot et al. 2008). Through the work of the Commission, and other related research, extensive evidence has been produced documenting the overwhelming impact of these determinants on health (Richmond et al. 2005) and the challenge of implementing actions to address health inequities.

There nevertheless remains a strong tendency for public health practitioners and traditional epidemiologists to concentrate on the influence of 'risk factors' related to living and working conditions of individuals, rather than to more systematically address the social dynamics that affect population health (Spiegel et al. 2015). In this regard, they risk being 'prisoners of the proximate' in the words of an acclaimed epidemiologist who emphasized the need for giving greater attention to contextual considerations (McMichael 1999). Lifestyle drift describes the tendency in public health to focus on individual behaviours such as smoking, diet, alcohol, and drugs but ignores the drivers of the causes of the causes (Marmot and Allen 2014). This approach sees that individuals are largely responsible for their own health and can improve their health through better health behaviours. However, this can be challenged by the numerous complex interactions and forces operating at the local, national and global scale that directly and indirectly influence health.

To more explicitly focus on the systemic circumstances that underlie the persistence of 'risk factor' inequalities among populations, an orientation that emphasizes 'processes of social determination' emerged in Latin America, a region in the Global South with a long history of pronounced income inequality (Breilh 2008). Applying such consideration of the social determination of health (rather than social determinants) calls for a more integrated examination of the broader scales and influences on health. These wider set of forces and systems alluded to in the WHO definition include economic policies and systems, development agendas, social policies and political systems. The WHO has also defined the social determinants of health as 'the unequal distribution of power, income, goods, and services, globally and nationally, the consequent unfairness in the immediate, visible circumstances of peoples' lives—their access to healthcare, schools, and education, their conditions of work and leisure, their homes, communities, town or cities—and their chances of leading a flourishing life' (Marmot et al. 2008).

This calls for a move beyond reductionist framings to more clearly discern the social processes and underlying power relationships that shape individual 'risk factor' social determinants. This allows us to first, re-conceptualize health as a complex, multidimensional object that is submitted to a process of determination and secondly, systematically address the social dynamics that affect population health (Spiegel et al. 2015). Typically, the biomedical model concentrates on the consequences of individuals' diets or lifestyle choices which are delineated as 'risk factor' determinants. A social determination orientation would encourage focus on the how the forces that are systemically present may be restricting options and choices at different scales, leading to limited opportunities for pursuing healthier alternatives—and better explain the persistence of patterns or options for pursuing different possible courses.

The example of nutrition transition in Central America can illustrate the difference between social determinants and social determination. In recent years, the rates of overweight and obesity as well as stunting have risen significantly in the region. Diet has been attributed to both phenomena as a result of major shifts in intake of less-healthful low-nutrient-density foods and sugary beverages, and changes in away-from-home eating. Similarly, 'sedenterization' of lifestyles has been also been purported by the nutrition literature. Yet nutrition transition in Central America is a dynamic, complex, and diachronic process.

An individual's socioeconomic status is often used a proxy for nutritional and health status and is commonly cited as a social determinant. In the past, low socioeconomic status was synonymous with poor nutrition and nutrient intake leading problems including stunting. More recently, low socioeconomic status has become associated with obesity (Dodd 2011). Deep inequalities are at the core of these ecologies. Particularly, unequal access to nutrients based on gender, age and ethnicity are pronounced throughout Central America. Thus, these inequalities, rooted in historical oppression, cultural traditions, or social norms [broadly conceptualized as social determination], manifest themselves in the bodies and health of marginalized individuals (Dodd 2011).

Political Ecology of Health

A society's economic, political, and social relationships affect both how people live and their ecological context, and, in doing so, shape patterns of disease and wellness distribution. Socio-political theories thus provide an alternative to biomedical approaches to understanding and addressing health concerns. They focus principally on power, politics, economics, and rights as key drivers of societal determinants of health. Such a perspective allows for multi-scalar analysis of the particular set of social relations, networks, and experiences that produce place.

In pursuing such focus on power relations at the individual, institutional, and global levels and the convergence of experiences in a given sociohistorical context and situational landscape, the political ecology orientation to health geography provides a particularly appropriate approach for generating a deeper understanding of the intersectionality of these influences (Krieger 2008) to guide how the challenge of 'global health' framings can be addressed (Brisbois et al. 2018).

It has been argued (Brisbois et al. 2018) that political ecology provides an orientation that is less a methodology than 'a community of practice and ... a certain kind of text' (Robbins 2012). This encompasses approaches applied to integrate political economy and disease ecology to distinguish a 'political ecology of health' approaches to others that distinguish a more holistic and dynamic 'political ecology of health' orientation to also take fuller account of 'long term well-being and individual agency' to understand geographies of human health. The framework guiding this approach is rooted in historical narratives of cultural and social factors in the analysis of health and provides a structural approach that attempts to connect large-scale political, social and economic processes to local health and well-being (Rocheleau 2008). Political ecology has thus emerged as a geographic subfield which examines the contextual realities of resource use and decision making. It can illustrate how health is embedded within social networks that increase vulnerability to diseases and shape health decision making and how health is shaped through the relationships between social and environmental systems (King 2009).

The use of place and landscape in the 'new' health geography offers a direct connection with political ecology since both of these subfields have shown a commitment to understanding the particular set of social relations, networks, and experiences that produce place (Jackson and Neely 2014). Political ecology provides the theoretical framework that makes power central to the analysis of the relationships between social and environmental systems. Understanding of diseases in 'places' is in the broader interest of identifying the world's most at risk regions in terms of the influence of social, political and behavioural changes on disease locales. A convergence of political geography and political ecology, perhaps, allows more nuanced and powerful explanations of ecological and social change.

The examination of power relationships can uncover inequities in decision-making at a variety of scales. This emphasizes health consequences of uneven development, institutional control over health status and potentially the structural changes and policies that affect disease hazards and human vulnerability. The discourse on the political ecologies of health through the exploration of large-scale social, economic and political influences include an interesting range of examples. Among the earliest was the relationship between large-scale development in Africa, and the transmission and propagation of diseases such as malaria and schistosomiasis, political structures on health and disease (Hunter 2003). Another example is the study of cholera in Mozambique through an ecologically based assessment of pathogens in places with assessment of the influence of vulnerability through displacement and structural changes associated with macro-economic policies (Williams et al. 2010).

Mapping Outcomes for Mothers [MOM]: A Case Example

Maternal health has traditionally taken a micro and individualistic approach to addressing maternal morbidity and mortality. The 'three delay model' introduced in the early 2000s sought to go beyond this by drawing much needed attention to the broader factors (Thaddeus and Maine 1994). These delays—delay in decision making; delay in reaching care; and delay in receiving adequate health care—took into consideration the societal context within which women live and the societal causes of maternal mortality. However, this explanatory model is often used to examine sociocultural 'risk factors' without attempting to consider the processes whereby culturally shared beliefs are integrated within the political and economic context.

In acknowledging the failure to meet the Millennium Development Goals maternal mortality targets (World Health Organization 2016), the maternal health agenda is now moving to include structural thinking and structural policy-based solutions. The 'Strategies Toward Ending Preventable Maternal Mortality' (EPMM) initiative was introduced to catalyze global action to eliminate wide disparities in the risk of death and end preventable maternal deaths (World Health Organization 2015). One of the guiding principles of EPMM calls on governments to 'address structural, historical and social determinants of health, and gender discrimination including economic inequality and workplace discrimination, to ensure equal outcomes for women and girls' (World Health Organization 2015).

In attempting to understand the structural processes that underlie the complex relationship between maternal health, environment and economy, we present a qualitative study that was undertaken in Mozambique to understand the community perspectives of the links between the social, environmental, cultural, economic and political processes that have simultaneously shaped maternal health—and consider how applying a political ecology lens to this experience can point to valuable lessons (Firoz et al. 2016).

Study Description

MOM was undertaken in rural southern Mozambique to develop a mobile health application (mhealth app) for mapping and predicting community specific risk of maternal morbidity and mortality. Using Geographic Information Systems (GIS) to overlay health with the local social and environmental contexts, it aims to elucidate the social and environmental factors associated with adverse maternal and perinatal outcomes. Ultimately, the hope is that it would strengthen the woman's ability to cope in the face of high risk by recognizing and communicating these risk factors that promote maternal resilience.

The first step in the development of MOM was to gather data on community perspectives of the determinants of maternal health to generate a list of predictors or variables, which we discuss below. Following this, we prioritized the variables through a Delphi consensus process, collected empirical data on the variables through a retrospective cohort study and then conducted spatial and statistical analyses to explore the association between these variables and adverse maternal outcomes (Makanga et al. 2019).

To identify key community groups' perspectives on the perceived determinants of maternal health in rural areas of southern Mozambique, we conducted an initial exploratory qualitative study (Firoz et al. 2016). A total of ten focus group discussions (FGD) were conducted with pregnant women, reproductive age women, matrons (elderly women in the community who serve as traditional birth attendants), male partners, community leaders, and health workers. The study area for focus groups consisted of four administrative posts, two in Gaza and two in Maputo province. To understand the context within which women live, in-depth interviews (IDI) were conducted with 12 administrative post chiefs in Gaza and Maputo.

Context

Mozambique has one of the highest rates of maternal mortality in sub-Saharan Africa with a maternal mortality ratio (MMR) of 489 per 100,000 live births (The World Bank 2015). One of the main drivers of the high MMR is a lack of health services, which were destroyed during years of war and have not yet been rebuilt

(Pfeiffer et al. 2017). Longstanding underfunding and the health budget cuts initiated as part of economic restructuring in 1987 have further left all Mozambican government health services compromised (Pfeiffer et al. 2017). In recent years, Mozambique has experienced a deepening economic and political crisis provoked by the accumulation of new debt and political instability (Pfeiffer et al. 2017). This has left the Ministry of Health with little ability to cope with the current national economic crisis and the consequences to the population health are expected to get worse (Cliff 1991; Pfeiffer et al. 2017).

Findings

A broad range of political, economic, sociocultural, and environmental determinants of maternal health were identified by community representatives.

Political Determinants

It was perceived that the civil war resulted in local unemployment and poverty that had a number of downstream effects. The lasting impact of the war included the loss of lives, infrastructure, and livestock leading to unemployment. While women, male partners, community leaders and health workers did not directly comment on the link between the war and maternal health, unemployment was consistently mentioned by women of reproductive age as an important factor influencing their well-being.

Economic Determinants

Women of reproductive age as well as health workers stated that financial constraints limited women's ability to access transport and care, especially to buy medication. At the facilities, women often incurred additional expenses beyond the cost of services, such as the purchase of food while in care, however, women were frequently unable to comply due to lack of funds. Women stated that if their partners could not assume the medical expenses during pregnancy or did not share their salaries, it could lead to vulnerability or pregnancy complications.

Sociocultural Determinants

A woman's marital status during pregnancy was identified as an important health determinant by women, health workers and community leaders. Women perceived single women (women whose partners abandoned them after finding out the pregnancy) as vulnerable particularly due to financial constraints. During focus group discussions, pregnant women described that single women have more complications because they had 'to fulfil all the requirements of the house' [Focus group discussion with pregnant women, Manhica].

Community leaders believed widows had the greatest chance of complications because they were poor. In contrast, health workers identified divorced women as at high risk, explaining that 'it affects [them] psychologically… it even causes trouble… because she thinks, I'm pregnant but because of [lack of] sustenance, what will be of that child tomorrow, if it happens to be born' [Focus group discussion with health workers, Chongoene, Gaza]. These respondents also felt that separated women were vulnerable because of a lack of support.

A perception shared by both women and community leaders was that relationships tended to change if women became sick during pregnancy. Reproductive age women, health workers and community leaders described that partners of sick women no longer wanted to take care of them and would at times abandon them.

Environmental Determinants

Poor access to roads and transport was identified as an environmental determinant that impacts maternal health. Male partners identified that women often walked long distances to reach facilities. Yet several study areas were described by administrative post chiefs as sandy and therefore, required large 4 × 4 vehicles for transport. Other areas were described as muddy or had potholes. An important consideration for accessing roads was seasonality, particularly the rainy season where many regions are prone to floods. When asked about transportation, participants mentioned that often they do not have access to vehicles, and that vehicles are used 'only if the person is serious'.

Discussion

The case example of MOM demonstrates that a reductionist approach does not capture the complex and dynamic interactions that result in the high MMR in Mozambique. A reductionist approach might suggest that rural women simply do not utilize maternal health care services and, therefore, the lack of care leads to pregnancy complications and high rates of maternal mortality. Using the lens of political ecology of health, we can unpack the local context and bring to the surface the historical context of the country ['place'] and the multi-scalar dimensions such as economic restructuring leading to political instability and civil war which in turn has resulted in poverty and gendered economic marginalization (Cliff 1991; Pfeiffer et al. 2017). Our interpretation is similar to that of Chapman who described that Mozambique's history of internal war had a profound impact on the societal structure and women's reproductive vulnerability in her ethnographic work (Chapman 2006).

The lasting impact of war has left the Mozambican government with little funds resulting in poor road infrastructure. This has been further compounded by the effects of natural disasters leaving roads muddy or with potholes. Seasonality especially heavy rains render roads inaccessible several times during the year. A study by our group using spatio-temporal modelling found that most women in our study area either walked or used public transport to access maternal care at the primary level, while most primary facilities provided transport to higher level facilities. A total of 13 of the 417 communities in the study area were completely isolated from maternal health services as a result of flooding at some time during the study timeline (Makanga et al. 2017). Accessing a vehicle, specifically a 4 × 4 vehicle, becomes necessary to traverse the roads and reach a facility. However, poverty resulting from unemployment due to the local effects of war leaves women and their families without money to access a vehicle and reach facilities.

A reflex response may be to institute more facilities to increase utilization without fully understanding the deeper reasons behind why women are not reaching facilities. Similar to the work by Atuoye et al. in Ghana (Atuoye et al. 2015), we find that the factors influencing transport access interact within a sociocultural, political, and physical environment to create a poor transport system resulting in low access to maternal health care services in rural communities. Therefore, an understanding of the local and contextual influences on health is necessary when developing strategies and interventions to address maternal morbidity and mortality in rural Mozambique. Participation and local community involvement is a necessary prerequisite to developing sustainable solutions. However, without systematic efforts to incorporate women's voices, needs, and interests, participatory approaches will only serve to suppress women's agency and reinforce existing gender inequalities (Adams et al. 2018).

Community-based participatory research (CBPR) can be a transformative paradigm to bridge the gap between science and practice through community engagement and social action to increase health equity. It allows for the potential to develop, implement, and disseminate effective interventions through strategies that redress power imbalances, facilitate mutual benefit between the community and academic partners and promote reciprocal knowledge translation, incorporating community theories into the research (Wallerstein and Duran 2010).

Based on the findings of MOM as identified through community-based research, we developed RoadMApp, which is further discussed in Chap. 14. RoadMApp is a location sensitive mobile technology solution built upon a novel spatio-temporal model of access to maternal health care (Makanga et al. 2017). RoadMApp accounts for road and weather conditions in near-real time, and links women to local transport resources that are available to facilitate their transit to maternal care. It also translates culturally established micro-savings traditions to finance this process.

By applying the political ecology of health framework and examining the intersections between health, geography, environment, political structures and maternal health, we find that it is imperative to involve stakeholders beyond the health sector.

For example, in order for successful uptake and implementation of RoadMApp, involving stakeholders like the Ministry of Transport, telecommunications organizations and local transport providers will be imperative. These stakeholders have a critical role in ensuring timely access to health care services for pregnant women.

Recently, a series on multi-sectoral collaboration featuring 12 country case studies has provided evidence for an emergent collaborative model for maternal and child health. It highlights evidence for the power of a 'learning society' that has continuous learning, active citizenship, and social well-being as its primary goals (Graham et al. 2018).

While multi-sectoral collaboration has relevance across diverse geographical, economic, social, cultural, and historical contexts, it is also dynamic and evolving. Stakeholders and their engagement change across different components and periods, highlighting the importance of realistic time frames, diverse evidence and ideas, and of 'learning and adapting while doing' to yield transformative results (Graham et al. 2018).

Conclusion

The inclusion of the social determinants of health in understanding health disparities show that geography and health are intrinsically linked. A society's economic, political, and social relationships affect both how people live and their ecological context, and, in doing so, shape patterns of disease distribution and ill health. Frameworks like political ecology of health challenge the narrow biomedical perspective and allow an examination of the multi-scalar social processes that influence health and drive health inequities. Adopting a language that focuses more directly on the social processes underlying health inequities can prompt a different approach to research and related policy and praxis. Action outside the health care sector can lead to more effective action to improve health and reduce inequities. Critical to this, is the inclusion of communities in the development of strategies and interventions to address disparities.

References

Adams, E. A., Juran, L., & Ajibade, I. (2018). 'Spaces of Exclusion' in community water governance: A feminist political ecology of gender and participation in Malawi's Urban Water User Associations. *Geoforum, 95*, 133–142. https://doi.org/10.1016/j.geoforum.2018.06.016.

Asthana, S., Curtis, S., Duncan, C., & Gould, M. (2002). Themes in British health geography at the end of the century: A review of published research 1998–2000. In *Selected Papers from the 9th International Symposium on Medical Geography* (Vol. 55(1), pp. 167–173). https://doi.org/10.1016/S0277-9536(01)00211-8

Atuoye, K. N., Dixon, J., Rishworth, A., Galaa, S. Z., Boamah, S. A., & Luginaah, I. (2015). Can she make it? Transportation barriers to accessing maternal and child health care services in rural Ghana. *BMC Health Services Research, 15*(1), 333. https://doi.org/10.1186/s12913-015-1005-y.

Breilh, J. (2008). Latin American critical ('Social') epidemiology: New settings for an old dream. *International Journal of Epidemiology, 37*(4), 745–750. https://doi.org/10.1093/ije/dyn135.

Brisbois, B. W., Harris, L., & Spiegel, J. M. (2018). Political ecologies of global health: Pesticide exposure in southwestern Ecuador's Banana Industry. *Antipode, 50*(1), 61–81. https://doi.org/10.1111/anti.12340.

Chapman, R. R. (2006). Chikotsa—Secrets, silence, and hiding: Social risk and reproductive vulnerability in Central Mozambique. *Medical Anthropology Quarterly, 20*(4), 487–515. https://doi.org/10.1525/maq.2006.20.4.487.

Cliff, J. L. (1991). The war on women in Mozambique: Health consequences of south African destabilization economic crisis and structural readjustment. In Women and Health, Africa ed. Meredith Turshen. 15-34. Trenton, NJ. Africa World Press.

Collins, A. E. (2002). Health ecology and environmental management in Mozambique. *South African Perspectives on Geography of Health, 8*(4), 263–272. https://doi.org/10.1016/S1353-8292(02)00005-9.

Dodd, W. (2011). Towards a political ecology of nutritional transitions of Central America: The construction of nutrition-deficient ecologies (Vol. 19, pp. 119–135). Totem.

Dummer, T. J. B. (2008). Health geography: Supporting public health policy and planning. *Canadian Medical Association Journal, 178*(9), 1177. https://doi.org/10.1503/cmaj.071783.

Firoz, T., Vidler, M., Makanga, P. T., Boene, H., Chiaú, R., Sevene, E., Magee, L. A., von Dadelszen, P., Munguambe, K., & The CLIP Working Group. (2016). Community perspectives on the determinants of maternal health in rural southern Mozambique: A qualitative study. *Reproductive Health, 13*(2), 112. https://doi.org/10.1186/s12978-016-0217-x.

Graham, W. J., Kuruvilla, S., Hinton, R., Veitch, E., & Simpson, P. J. (2018). Multisectoral collaboration for health and sustainable development. *BMJ, 363*, k4868. https://doi.org/10.1136/bmj.k4868.

Hoey, B. A. (2007). Therapeutic uses of place in the intentional space of purposive community. In A. Williams (Ed.), *Therapeutic landscapes: Advances and applications*. Aldershot: Ashgate.

Hunter, J. M. (2003). Inherited burden of disease: Agricultural dams and the persistence of bloody urine (Schistosomiasis hematobium) in the Upper East Region of Ghana, 1959–1997. *Social Science & Medicine, 56*(2), 219–234. https://doi.org/10.1016/S0277-9536(02)00021-7.

Jackson, P., & Neely, A. H. (2014). Triangulating health: Toward a practice of a political ecology of health. *Progress in Human Geography, 39*(1), 47–64. https://doi.org/10.1177/0309132513518832.

King, B. (2009). Political ecologies of health. *Progress in Human Geography, 34*(1), 38–55. https://doi.org/10.1177/0309132509338642.

Krieger, N. (2008). Proximal, distal, and the politics of causation: What's level got to do with it? *American Journal of Public Health, 98*(2), 221–230. https://doi.org/10.2105/AJPH.2007.111278.

Makanga, P. T., Schuurman, N., Sacoor, C., Boene, H. E., Vilanculo, F., Vidler, M., Magee, L., von Dadelszen, P., Sevene, E., Munguambe, K., & Firoz, T. (2017). Seasonal variation in geographical access to maternal health services in regions of southern Mozambique. *International Journal of Health Geographics, 16*(1), 1. https://doi.org/10.1186/s12942-016-0074-4.

Makanga, P. T., Sacoor, C., Schuurman, N., Lee, T., Vilanculo, F. C., Munguambe, K., Boene, H., Ukah, U. V., Vidler, M., Magee, L. A., Sevene, E., von Dadelszen, P., & Firoz, T. (2019). Place-specific factors associated with adverse maternal and perinatal outcomes in southern Mozambique: A retrospective cohort study. *BMJ Open, 9*(2), e024042. https://doi.org/10.1136/bmjopen-2018-024042.

Mansfield, B. (2008). Health as a nature—Society question. *Environment and Planning A: Economy and Space, 40*(5), 1015–1019. https://doi.org/10.1068/a40366.

March, D., & Susser, E. (2006). The eco- in eco-epidemiology. *International Journal of Epidemiology, 35*(6), 1379–1383. https://doi.org/10.1093/ije/dyl249.

Marmot, M., & Allen, J. J. (2014). Social determinants of health equity. *American Journal of Public Health, 104*(S4), S517–S519. https://doi.org/10.2105/AJPH.2014.302200.

Marmot, M., Friel, S., Bell, R., Houweling, T. A. J., & Taylor, S. (2008). Closing the gap in a generation: Health equity through action on the social determinants of health. *The Lancet, 372*(9650), 1661–1669. https://doi.org/10.1016/s0140-6736(08)61690-6.

McMichael, A. J. (1999). Prisoners of the proximate: Loosening the constraints on epidemiology in an age of change. *American Journal of Epidemiology, 149*(10), 887–897. https://doi.org/10.1093/oxfordjournals.aje.a009732.

Pfeiffer, J., Gimbel, S., Chilundo, B., Gloyd, S., Chapman, R., & Sherr, K. (2017). Austerity and the "sector-wide approach" to health: The Mozambique experience. *Social Science & Medicine, 187*, 208–216. https://doi.org/10.1016/j.socscimed.2017.05.008.

Richmond, C., Elliott, S. J., Matthews, R., & Elliott, B. (2005). The political ecology of health: Perceptions of environment, economy, health and well-being among 'Namgis First Nation'. *Special Section: Therapeutic Landscapes: An Evolving Theme, 11*(4), 349–365. https://doi.org/10.1016/j.healthplace.2004.04.003.

Robbins, P. (2012). *Political ecology: A critical introduction* (2nd ed.). Chichester: John Wiley & Sons.

Rocheleau, D. E. (2008). Political ecology in the key of policy: From chains of explanation to webs of relation. *Conversations Across the Divide, 39*(2), 716–727. https://doi.org/10.1016/j.geoforum.2007.02.005.

Spiegel, J. M., Breilh, J., & Yassi, A. (2015). Why language matters: Insights and challenges in applying a social determination of health approach in a north-south collaborative research program. *Globalization and Health, 11*(1), 9. https://doi.org/10.1186/s12992-015-0091-2.

Thaddeus, S., & Maine, D. (1994). Too far to walk: Maternal mortality in context. *Social Science & Medicine, 38*(8), 1091–1110. https://doi.org/10.1016/0277-9536(94)90226-7.

The World Bank. (2015). *Maternal mortality ratio*. Retrieved from https://data.worldbank.org/indicator/sh.sta.mmrt.

Wallerstein, N., & Duran, B. (2010). Community-based participatory research contributions to intervention research: The intersection of science and practice to improve health equity. *American Journal of Public Health, 100*(S1), S40–S46. https://doi.org/10.2105/AJPH.2009.184036.

Williams, L., Collins, A. E., Bauaze, A., & Edgeworth, R. (2010). The role of risk perception in reducing cholera vulnerability. *Risk Management, 12*(3), 163–184. https://doi.org/10.1057/rm.2010.1.

World Health Organization. (2015). *Strategies toward ending preventable maternal mortality (EPMM)*. Geneva: World Health Organization.

World Health Organization. (2016). *Social determinants of health*. Geneva: World Health Organization. Retrieved from www.who.int/topics/social_determinants/en/.

Part III
Geo-Enabling Health Decisions

Chapter 11
Geo-enabled Trauma Registries: The Case of Cape Town, South Africa

Germaine Tuyisenge, Leah Rosenkrantz, Nadine Schuurman, Ellen Randall, and Morad Hameed

Abstract Trauma registries are repositories of systematically collected clinical, demographic and epidemiological data on patients who have sustained injuries (Nwomeh et al., World J Emerg Surg 1(1):32, 2006). With an example of a trauma hospital in Cape Town, South Africa, this chapter highlights the story of a decade-long effort to develop a trauma registry. Contained within this account are many sub-plots, including the value of trauma registries in the effort to improve injury prevention and management, the obstacles to the complete and permanent implementation of registries in low- and middle-income settings, the transition from paper to digital registry records and finally, the many geospatial applications of a detailed trauma registry.

Keywords Trauma surveillance · Geo-enabled decisions · Injury prevention policy · Trauma registries

Introduction

The Burden of Injury

In 2014, the World Health Organization (WHO) reported that nearly 6 million people die of injury-related causes worldwide annually—more than 1.5 times the number of deaths resulting from HIV/AIDS, malaria and tuberculosis combined. Injuries primarily result from acts of violence (interpersonal or to oneself), motor vehicle

G. Tuyisenge (✉) · L. Rosenkrantz · N. Schuurman
Department of Geography, Simon Fraser University, Burnaby, BC, Canada

E. Randall
School of Population and Public Health, University of British Columbia, Vancouver, BC, Canada

M. Hameed
Division of General Surgery, Vancouver General Hospital, Vancouver, BC, Canada

Department of Surgery, University of British Columbia, Vancouver, BC, Canada

P. T. Makanga (ed.), *Practicing Health Geography*, Global Perspectives on Health Geography, https://doi.org/10.1007/978-3-030-63471-1_11

collisions, burns, drownings, falls and poisonings (World Health Organization (WHO) 2014a).

While injury is a truly global health issue, a great disparity exists between prevalence and mortality rates between low- and middle-income countries (LMICs) and high-income countries (HICs). In fact, more than 90% of all deaths from injuries occur in LMICs (World Health Organization (WHO) 2014a). Furthermore, those with life-threatening but treatable injuries in an LMIC are more than six times as likely to die as someone with similar injuries in an HIC (Mock et al. 1998). The reality underlying these numbers is multifactorial, but it can largely be attributed to the infrastructural, financial and human resource scarcity in LMICs' health care systems, which limit their ability to provide timely, adequate care and ongoing management to the injured (Mock et al. 1993). Additionally, a considerable number of trauma cases fail to even reach trauma centres because of issues with spatial access to these facilities and/or with inadequate referral systems between primary healthcare centres and advanced level trauma centres in most LMICs (Govender et al. 2012; Mock et al. 1993; Nicol et al. 2014).

Trauma Registries: A Brief History

Trauma registries are databases that document fields of information related to the epidemiology of injury and process of care delivered to trauma patients presenting to hospitals or health centres (Nwomeh et al. 2006). The data collected can be used for several purposes, but are primarily used for injury surveillance to inform preventative action, to improve quality of trauma care and as a benchmarking tool against other hospitals, regions or even nations depending on the health system level at which the registry operates (Moore and Clark 2008). The databases typically include variables for patient demographics, epidemiologic information—e.g. cause, nature and severity of the injury—as well as pre-hospital and hospital care, and outcomes.

For over three decades, trauma registries have been a key part of trauma systems and have been used to support trauma management and research (Nwomeh et al. 2006). In the United States, trauma registries emerged in the late 1960s as institution-based tools for trauma management and were developed based on trauma centres' needs and available resources (Zehtabchi et al. 2011). In fact, the first ever computerized trauma registry was established in 1969 at the Cook County Hospital in Chicago, Illinois, paving the way for many more hospitals in the state and rest of the country to follow suit (Nwomeh et al. 2006). The registry was even used as the prototype for the Illinois Trauma Registry which was established in 1971—one of the first multi-hospital databases pooling data from 50 trauma centres across the state (Nwomeh et al. 2006). Since then, numerous other hospitals and countries have followed suit, implementing their own registries, many of which are regional-, state-, or even nation-wide. A scoping review published in 2012 identified 293 unique trauma registries across the globe from an extensive search of academic literature published on these registries, and this number is growing (O'Reilly et al. 2012).

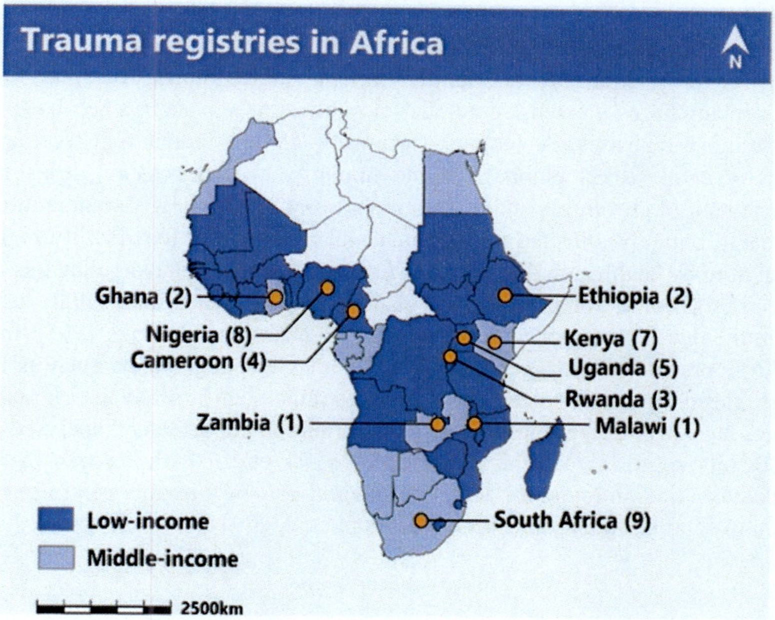

Fig. 11.1 Map of trauma registries in Africa

However, the number of trauma registries in Africa are still low (Rosenkrantz et al. 2019). Figure 11.1 is a map of trauma registries on the continent.

The Many Functions of a Trauma Registry

As mentioned previously, trauma registries can serve multiple purposes. Research using the data from registries typically focusses on injury surveillance, improving quality of trauma care and internal or external benchmarking of trauma care (Moore and Clark 2008; O'Reilly et al. 2013; Rosenkrantz et al. 2019). More recently, trauma registry data have been used to assess geospatial patterns of injury and this can be seen as a critical component of injury surveillance (Chow et al. 2012; Cinnamon and Schuurman 2010).

Injury Surveillance

Systematically collected trauma registry data can be analysed and interpreted for injury surveillance purposes with the aim of providing useful data about the distribution and determinants of injury to inform prevention strategies and better improve resource allocation. In countries where trauma registries have been integrated as a component of trauma management—typically HICs (Nwomeh et al. 2006)—they have proved a powerful tool in the identification and monitoring of high-risk populations (O'Reilly et al. 2012). Countries like the United States, Australia, Canada and Germany have an extensive number of publications on trauma because of data from trauma registries. A scoping review on global trauma registry mapping (O'Reilly et al. 2012) emphasized the importance of trauma registries in the advancement of prevention and management of injuries. In HICs, trauma information that is routinely collected at different health system levels (e.g. facility, regional) is then used by health care decision makers in establishing related policy (Nwomeh et al. 2006). For example, trauma registry data have been successfully used to inform the development of targeted preventive measures (Rutledge 1995). In fact, the development and enforcement of traffic rules and regulations such as motor vehicle speed control and mandatory use of restraints such as seatbelts, car seats and helmets have been heavily influenced by information collected and analysed about injuries relating to motor vehicle collisions (Mock et al. 2005; Toroyan 2009). In this context, trauma registries act as an integral part of a trauma centre or health system to support injury prevention (Racioppi et al. 2004).

Spatial Analysis of Injuries

From a geographical perspective, trauma registries are an important tool for delineating the spatial distribution of trauma incidents—a key input to injury surveillance and how referral systems facilitate the timely management of trauma cases (Mackenzie and Fowler 2003). To advance trauma surveillance, the WHO recommends that a minimum dataset on trauma patients should be collected and processed on an ongoing basis, including spatial elements such as location where the injury was sustained (Holder et al. 2001). Trauma registry data can be used to assist in the mapping of trauma incidents, providing key information about the relationship of place to injury.

Studies focusing on the spatial aspects of trauma have made important contributions to shedding light on the relationships between geographic information and other correlates of injury (Schuurman et al. 2008). A geospatial analysis of where injuries were originally sustained in the United States between 2005 and 2007 (Newgard et al. 2011) included an evaluation of the socio-economic status of neighbourhoods with a high incidence of trauma across the country. The results indicate that incidence of injury was higher in lower socio-economic neighbourhoods based on measures such as unemployment and income levels. Similarly, using the British Columbia Trauma Registry data, researchers and trauma surgeons in Vancouver, Canada, analysed violent trauma incidents that occurred in Vancouver between 2001 and 2008 (Walker et al. 2014). Using Geographic Information Systems (GIS) to map where and when injury arising from interpersonal violence happened, the resulting analysis provided a map of high-risk areas that allowed further exploration into the relationships between the characteristics of these areas and other correlates of injury.

Quality Improvement

Quality improvement focusses on the implementation and measurement of corrective actions to better patient care (Mock 2009). Trauma registries provide critical longitudinal data for quality improvement programmes, which are important to supporting optimal injury care provision and management at the level of trauma care centres themselves, as well as at regional and/or national health system levels (Cameron et al. 2005; Moore and Clark 2008). One example of the use of trauma registry data for this purpose is illustrated in a study conducted in Hong Kong that sought to evaluate whether trauma team activation based on a 10-point protocol is associated with better survival (Rainer et al. 2007). Using data from the hospital's trauma registry, the authors were able to show that while trauma team activation does not guarantee improved survival, compliance with activation protocols did optimize the process of care—a fact that may translate to improved survival in the long run (Rainer et al. 2007). Trauma registries in many other hospitals and at different health system levels have reported similar applications, using the data collected to critically evaluate outcomes of newly implemented measures, and allowing for improvements in all facets of trauma management (Cameron et al. 2005; Parreira et al. 2015).

Benchmarking of Trauma Care

Quality control relies on registry data to allow for routine assessment of key standardized indicators to benchmark the quality of care both within facilities over time, and across regional trauma centres (Mock 2009). These indicators include the available resources at each level of trauma care, such as human and financial resources, as well as measures of clinical care. They may vary according to the trauma services offered as well as the available health infrastructure (Mock et al. 2005). The standardization of data collection imposed by most trauma registries enables trauma centres to share and compare data with the aim of better understanding where improvements in care and care management are needed, and how high-performing centres differ from those who whose performance is lower on key indicators (Cameron et al. 2005; Owen et al. 1999). These improvements can range from increasing human or other resources at a given centre (Pino Sánchez et al. 2015) to improving regional referral systems so that patients are directed to the most appropriate level of trauma centre to receive the care needed (Cameron et al. 2005; Owor and Kobusingye 2001).

Trauma Registries in Low- and Middle-Income Countries (LMICs)

Barriers to Trauma Registry Implementation in LMICs

While recent years have shown a surge in trauma registries in LMICs, trauma registries for the most part have been largely confined to hospitals in HICs (Rosenkrantz et al. 2019). This trauma registry gap poses significant barriers to reducing the prevalence of injury in LMICs, as well as improving the quality of trauma care delivered to millions of people in these areas. Despite the overwhelming surge of trauma registries over the past five decades (Rosenkrantz et al. 2019), many variations in trauma registry design and implementation persist. While significant efforts have been made to standardize this process (O'Reilly et al. 2013), discrepancies still exist across countries (and in many instances, even within countries) about defining who qualifies as a trauma patient, and what information should be collected on those who are classified as such (Chow et al. 2012; Nwomeh et al. 2006; O'Reilly et al. 2013; Rosenkrantz et al. 2019). For example, trauma registries generally collect information on individuals presenting to trauma facilities with injuries; however, different trauma registries often have different views on what data are important to them. As such, some registries constrain their definition and only collect information on those trauma patients who are actually admitted to the hospital, whereas others have chosen to expand their definition and collect information even on those who die before making it to the emergency room, either on scene or en route to the hospital (Rutledge 1995).

During the 1990s, meetings on injury surveillance led by the WHO reached the conclusion that an internationally recognized and standardized classification of external causes of injury is necessary to improve injury research and benchmarking and to create consensus in describing these external causes (Scott et al. 2006). This resulted in the development of the International Classification of External Causes of Injury (ICECI), with first drafts released between 1998 and 2001 (Scott et al. 2006). The ICECI enables researchers and clinicians to describe, evaluate and monitor injuries, and provides a system for properly recording the information on the circumstances of the injury (World Health Organization (WHO) 2014b). Depending on the regional context of each trauma centre or the nature of the trauma care level, the ICECI structure may be internally adjusted to allow trauma registries to capture information that is most relevant to their needs for trauma prevention, management and research if it is conventionally approved by concerned players in trauma surveillance (Scott et al. 2006). The ICECI has served as an important tool for harmonizing the classification of injuries and is currently used by many trauma registries across the globe (Scott et al. 2006). It is important to remember that many other aspects of the trauma registry still remain unstandardized, such as how to define a trauma patient and what variables are necessary to collect. By working towards reaching an international consensus on these matters, we can further improve international benchmarking and trauma research.

Several other barriers exist that inhibit the implementation of trauma registries in LMICs, including the limited human resources of hospitals and a stiff competition for healthcare dollars with other more widely recognized diseases, like HIV or malaria (Nwomeh et al. 2006; O'Reilly et al. 2016). Further exacerbating these limitations is the sheer number of trauma cases seen by hospitals in LMICs, owing to the vast injury burden in these countries (Gosselin 2009). Another barrier often encountered by hospitals trying to establish their own trauma registry in LMICs is a lack of infrastructure, ranging from a deficit of computers and data storage facilities to issues with power and internet connectivity (Nwomeh et al. 2006). In addition to the lack of standardization described above, these constraints have made the implementation and operation of trauma registries in LMICs quite challenging.

Innovative Strategies for Implementing Trauma Registries

Innovative strategies such as the use of minimum datasets, the development of context-appropriate injury severity scoring measures and electronic data collection have allowed some hospitals to overcome these challenges (Nwomeh et al. 2006). For example, to address the human resource constraints and the timely nature of collecting numerous fields of data, pilot studies and hospital registries have explored the use of a minimum dataset that would reduce this burden by strategically collecting the most imperative variables for research purposes as defined by the hospitals themselves (Kobusingye and Lett 2000; Schuurman et al. 2011).

The use of context-appropriate injury severity scores has also been explored as a means to predict the outcome of injury depending on its severity, and to support appropriate management (Lefering 2009). The Kampala Trauma Score (KTS) is an example of a trauma registry tool specifically designed for a low-resource health setting (Haac et al. 2015). Developed in Kampala, Uganda and now used in trauma centres across Uganda and Ethiopia, the KTS is an injury severity scoring method—a simplified hybrid of the Revised Trauma Score (RTS) and the Injury Severity Score (ISS)—used for predicting mortality and need for admission (Haac et al. 2015; MacLeod et al. 2003; Weeks et al. 2014). Due to its simplicity, the KTS has the potential to ease trauma registry burden for health professionals in busy and under-resourced settings since it can be performed easily without needing to conduct a retrospective review of injury, as is required for other common scoring systems (MacLeod et al. 2003).

Another innovation that has been used to overcome some of the barriers present in low-resource settings is the use of electronic data collection—as opposed to paper-based charting—in operating the trauma registry. Use of electronic charting has been shown to be a more efficient form of data collection compared to manual data entry as it helps to cut down on time for clinicians to record data and eliminates the need to decipher others' handwriting (Häyrinen et al. 2008). It has also been shown to be statistically superior in both quality of data and field completion rates (Rosenkrantz et al. 2020). While the up-front cost of implementing an electronic data collection system is substantial, rapid developments in information technology and the availability of open-source software have brought the possibility of adopting this strategy within reach for many lower-resource registries (Fraser et al. 2005), including those in Pakistan, South Africa and Kenya (Mehmood et al. 2013; Zargaran et al. 2014).

Developing a Trauma Registry in Cape Town, South Africa

Injury in South Africa

South Africa, a middle-income country, serves as an example of how progress towards integrating trauma registries in the country's health system has been slow but critical for a country with a high incidence of trauma (Norman et al. 2007). Studies on trauma in South Africa have shown that the leading causes of injury are road-traffic collisions (RTCs) and interpersonal violence (Nicol et al. 2014). In 2009 alone, more than 50,000 trauma deaths were reported across the country (Matzopoulos et al. 2015) and the most common injury category was RTCs. A prospective study looking at 2011–2012 RTC data at Edendale hospital, KwaZulu-Natal, showed that over a period of only 10 weeks, 305 patients were seen at the hospital and 45 of these trauma cases resulted in fatalities (Parkinson et al. 2013). Furthermore, more than 14,000 RTC deaths were reported in South Africa in 2016 alone, translating into 25.2% of deaths per 100,000 population (International Transport Forum 2017).

Interpersonal violence is another major cause of injury in South Africa. After decades of apartheid, state-sponsored oppression and a protracted liberation struggle to achieve a non-racial democracy, violence has been woven into South Africa's complex history. Although political conflict has receded, violence remains a widespread strategy to resolve interpersonal conflict. Consequently, South Africa faces one of the highest burdens of intentional injury worldwide with an injury mortality rate of 87.8 per 100,000 population—higher than both the African continental average of 60.3 per 100,000 and over three times that of the global average of 28.7 per 100,000 (Matzopoulos et al. 2004). Given that South Africa is a country not at war, this high rate of intentional injury—which includes incidences of self-harm, collective violence and interpersonal violence in its definition—is alarming.

However, injury is not South Africa's only issue. In fact, the country faces what is known as a quadruple burden of disease, which includes (1) interpersonal violence and injury; (2) poverty-related, non-communicable diseases; (3) emerging chronic diseases; and (4) HIV/AIDS (Bradshaw et al. 2003). The constellation of these four public health issues amounts to a serious public health crisis as each demand significant human and financial resources beyond the government and health care system's capabilities. While it is clear that trauma registries would prove critical in hospitals here, as both a means of informing preventative action and improving quality of trauma care, this competition for human and financial resources has made the implementation of trauma registries especially challenging. Nonetheless, using strategies to overcome many of these barriers, over the last two decades several hospitals in the country have shown that trauma registry development is possible. The following case study documents the experience of one of these hospitals and the strategies they used to overcome these many barriers.

A Case Study of Groote Schuur Hospital in Cape Town, South Africa

Groote Schuur Hospital (GSH) is a leading, level 1 trauma centre in Cape Town, South Africa. The hospital was established in 1938 as a government-funded hospital and is currently the chief academic hospital for the University of Cape Town. GSH provides tertiary trauma care across the region and serves a metropolitan population of over 2 million people (Navsaria et al. 2009). As of 2009, the hospital had a total of 525 doctors and admitted over 43,000 patients in 2008, an estimated 12,000 of which were trauma patients (Nicol et al. 2014). During the same year, the Cape Town Trauma Registry (CTTR) was developed (Schuurman et al. 2011), an exercise that served as a pilot study that was used to assess the feasibility of such an endeavour. Following the initial success of CTTR came the development and implementation of the electronic Trauma Health Record (eTHR) in 2012 (Zargaran et al. 2014).

Development of the Cape Town Trauma Registry

In 2008, the CTTR was piloted using data from GSH trauma records for over 700 patients who presented at the trauma unit in a single month in the same year (Schuurman et al. 2011). The goal of this exercise was to develop an extensive, user-friendly and comprehensive trauma unit-based database to facilitate injury surveillance and comply with the major facets of international injury surveillance (Schuurman et al. 2011). In addition, developing such a tool was deemed a simplified yet comprehensive way to facilitate the trauma registry process and its integration in the trauma management system (Schuurman et al. 2011).

Prior to initiating the pilot study, a needs and feasibility assessment was conducted with hospital staff and clinicians to identify local hospital-based requirements, in addition to a review of both surveillance literature and the mandates of trauma registrars in Europe and North America. Prior to the pilot and in order to ensure richness of collected information, some of the commonly used international trauma scores (the KTS, the ISS and the Abbreviated Injury Score) were calculated in consultation with WHO guidelines (Schuurman et al. 2011). The pilot consisted of collecting and analysing demographic and epidemiological data of trauma patients who presented at GSH in October 2008. To preserve diversity among trauma cases and for patients' demographics to support the analysis, all trauma patients were included in data collection. This information was documented along with notes from nurses, ambulance forms and patient information which included history, diagnosis and treatment. A large dataset resulted from this pilot and was used to develop the CTTR form. In addition, data analysis proved that a minimum dataset could be used to calculate the ISS and other commonly used injury scores.

Moreover, one of the unique aspects of the CTTR is its ability to capture the spatial distribution of injury by collecting data on geographical location of trauma, which has helped guide injury prevention through improved understanding of how place and space influence injury prevalence and the varying socio-economic risk factors associated with those places (Schuurman et al. 2011; Zavalaa et al. 2008). To confirm the usability of a simplified trauma registry tool and test trauma surveillance, the CTTR was used for all trauma admissions between October 2010 and September 2011 at GSH (Nicol et al. 2014). The findings from this study show that young adult males were more likely to be involved in interpersonal violence leading to injury compared to older males and female counterparts. The results highlight that alcohol/substance abuse was involved in most of the registered trauma cases. In addition, the study revealed that most trauma cases occurred on the streets (48.6%), followed by homes (33.6%) and bars (4.8%). Most occurred in the townships of Cape Town (Nicol et al. 2014). This study confirmed that the information collected through the CTTR can help to strengthen prevention of trauma in the region. With the usability of the CTTR affirmed, the GSH was called upon to mobilize the necessary resources to integrate the trauma registry into its trauma management system in order to strengthen the quality of data collected through trauma admission records.

Development of the Electronic Trauma Health Record (eTHR)

In 2012, a team led by Dr. Morad Hameed, a trauma surgeon from Vancouver General Hospital and researcher at the University of British Columbia, Canada, developed the electronic Trauma Health Record (eTHR) to improve the efficiency of trauma data collection at GSH (Zargaran et al. 2014). By using the experience and observations gleaned from the CTTR, and through a consensus of trauma surgeons, surgical residents, nurses, data ethnographers and software designers, the specific features and capabilities of the eTHR were defined (Zargaran et al. 2018). The tablet-based application collects information on trauma patient admission, care process and discharge. The development of this application was deemed imperative as a means of improving the quality of data collected over the one-page patient admission record while also making injury data easily accessible for interpretation, sharing and analysis. The data gathered by the eTHR includes data from patient admission records plus information on injury prevention, care quality and procedures (Zargaran et al. 2014). During the pilot study of the application, a few

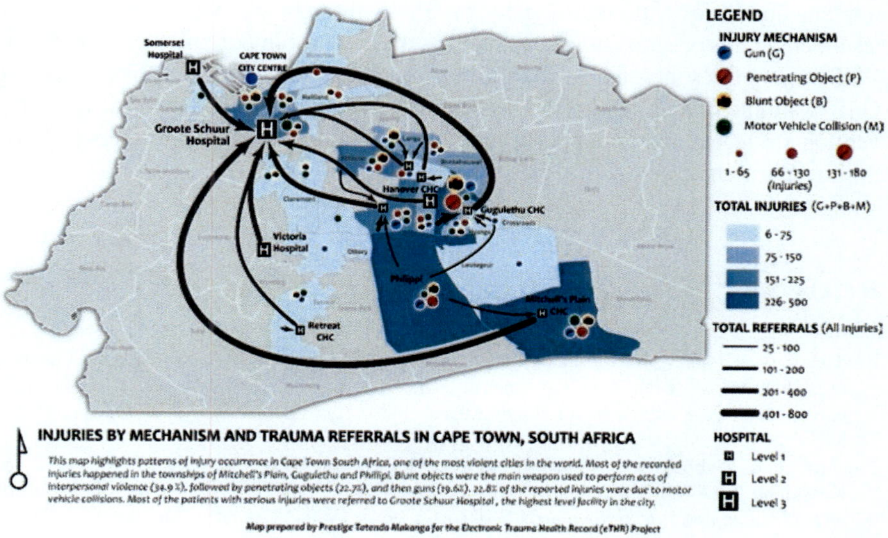

Fig. 11.2 Injury by mechanism and trauma referrals in Cape Town, South Africa

limitations to its use were revealed, including frequent electricity cut-offs and slow internet connectivity, which can be typical occurrences in hospitals and health facilities in LMICs. Nonetheless, the usability of the eTHR outweighed its limitations, an assessment supported by the volume of data collected and processed in a single month in 2013 as part of a test of the tool. This basic, easy-to-use instrument proved capable of providing relevant data to support both trauma care and injury surveillance, with minimal disruptions to the health facility workflow. Additionally, its ability to collect geospatial data opens up many more possibilities for research, including but not limited to prevention, access to care, and health inequalities. If implemented in LMIC trauma care settings, a tool such as eTHR could be a great benefit to trauma research and surveillance and also facilitate data-sharing between different levels of trauma management for patient record and referral purposes. Maps such as those in Fig. 11.2 illustrate how the data from eTHR could be used towards programme improvement by understand spatial patterns in injury occurrence, as well as how volumes of trauma traffic through the health facility referral system look like.

Conclusion

Trauma registries have been shown to improve injury surveillance and quality of trauma care. While widely used across hospitals in HICs, they are less common in LMICs due to financial and human resource barriers. Nonetheless, efforts to overcome these resource challenges have proven successful. The development and implementation of a trauma registry at GSH—through its many iterations—is a testament to this progress, and has served as the basis for important trauma research. While injury rates in South Africa still remain high, trauma registries like the one at GSH continue to help reduce the burden of injury by collating data critical to prevention strategies and to support improved morbidity and mortality rates through improved trauma care.

References

Bradshaw, D., Groenewald, P., Laubscher, R., Nannan, N., Nojilana, B., Norman, R., Pieterse, D., Schneider, M., Bourne, D. E., Timæus, I. M., Dorrington, R., & Johnson, L. (2003). Initial burden of disease estimates for South Africa, 2000. *South African Medical Journal, 93*(9), 682–688. https://doi.org/10.1016/j.trstmh.2008.07.002.

Cameron, P. A., Gabbe, B. J., McNeil, J. J., Finch, C. F., Smith, K. L., Cooper, D. J., Judson, R., & Kossmann, T. (2005). The trauma registry as a statewide quality improvement tool. *The Journal of Trauma: Injury, Infection, and Critical Care, 59*(6), 1469–1476. https://doi.org/10.1097/01.ta.0000198350.15936.a1.

Chow, C. B., Leung, M., Lai, A., Chow, Y. H., Chung, J., Tong, K. M., & Lit, A. (2012). Development of an electronic emergency department-based geo-information injury surveillance system in Hong Kong. *Injury, 43*(6), 739–748. https://doi.org/10.1016/j.injury.2011.08.008.

Cinnamon, J., & Schuurman, N. (2010). Injury surveillance in low-resource settings using geospatial and social web technologies. *International Journal of Health Geographics [Electronic Resource], 9*, 25. https://doi.org/10.1186/1476-072X-9-25.

Fraser, H., Biondich, P., Moodley, D., Choi, S., Mamlin, B., & Szolovits, P. (2005). Implementing electronic medical record systems in developing countries. *Journal of Innovation in Health Informatics, 13*(2), 83–95.

Gosselin, R. A. (2009). The increasing burden of injuries in developing countries: Direct and indirect consequences. *Techniques in Orthopaedics, 24*(4), 230–232. https://doi.org/10.1097/BTO.0b013e3181bfd56c.

Govender, I., Matzopoulos, R., Makanga, P., & Corrigall, J. (2012). Piloting a trauma surveillance tool for primary healthcare emergency centres. *South African Medical Journal, 102*(5), 303–306. https://doi.org/10.7196/samj.5293.

Haac, B., Varela, C., Geyer, A., Cairns, B., & Charles, A. (2015). The utility of the Kampala trauma score as a triage tool in a sub-Saharan African trauma cohort. *World Journal of Surgery, 39*(2), 356–362. https://doi.org/10.1007/s00268-014-2830-6.

Häyrinen, K., Saranto, K., & Nykänen, P. (2008). Definition, structure, content, use and impacts of electronic health records: A review of the research literature. *International Journal of Medical Informatics, 77*(5), 291–304. https://doi.org/10.1016/j.ijmedinf.2007.09.001.

Holder, Y., Peden, M., Krug, E. G., Lund, J., Gururaj, G., Kobusingye, O., & World Health Organization. (2001). *Injury surveillance guidelines*. Atlanta, GA: World Health Organization.

ITF (2017), Road Safety Annual Report 2017, OECD Publishing, Paris.

Kobusingye, O. C., & Lett, R. R. (2000). Hospital-based trauma registries in Uganda. *Journal of Trauma and Acute Care Surgery, 48*(3), 498.

Lefering, R. (2009). Development and validation of the revised injury severity classification score for severely injured patients. *European Journal of Trauma and Emergency Surgery, 35*(5), 437–447. https://doi.org/10.1007/s00068-009-9122-0.

Mackenzie, E. J., & Fowler, C. J. (2003). Epidemiology of injury. In *Trauma* (5th ed.). New York: McGraw-Hill Companies.

MacLeod, J. B. A., Kobusingye, O., Frost, C., Lett, R., Kirya, F., & Shulman, C. (2003). A comparison of the Kampala trauma score (KTS) with the revised trauma score (RTS), injury severity score (ISS) and the TRISS method in a Ugandan trauma registry: Is equal performance achieved with fewer resources? *European Journal of Trauma, 29*(6), 392–398. https://doi.org/10.1007/s00068-003-1277-5.

Matzopoulos, R., Norman, R., & Bradshaw, D. (2004). *The burden of injury in South Africa: Fatal injury trends and international comparisons*. Tygerberg: Medical Research Council-University of South Africa, Crime, Violence and Injury Lead Programme.

Matzopoulos, R., Prinsloo, M., Wyk, V. P., Gwebushe, N., Mathews, S., Martin, L. J., Laubscher, R., Abrahams, N., Msemburi, W., Lombard, C., & Bradshaw, D. (2015). Injury-related mortality in South Africa: A retrospective descriptive study of postmortem investigations. *Bulletin of the World Health Organization, 93*(5), 303–313. https://doi.org/10.2471/BLT.14.145771.

Mehmood, A., Razzak, J. A., Kabir, S., MacKenzie, E. J., & Hyder, A. A. (2013). Development and pilot implementation of a locally developed trauma registry: Lessons learnt in a low-income country. *BMC Emergency Medicine, 13*(1), 4.

Mock, C. (2009). WHO releases guidelines for trauma quality improvement programmes. *Injury Prevention, 15*(5), 359.

Mock, C. N., Adzotor, K. E., Conklin, E., Denno, D. M., & Jurkovich, G. J. (1993). Trauma outcomes in the rural developing world: Comparison with an urban level I trauma center. *The Journal of Trauma, 35*(4), 518–523.

Mock, C. N., Jurkovich, G. J., & nii-Amon-Kotei, D., Arreola-Risa, C., & Maier, R. V. (1998). Trauma mortality patterns in three nations at different economic levels: Implications for global trauma system development. *Journal of Trauma and Acute Care Surgery, 44*(5), 804.

Mock, C., Kobusingye, O., Anh, L. V., Afukaar, F., & Arreola-Risa, C. (2005). Human resources for the control of road traffic injury. *Bulletin of the World Health Organization, 83*, 294–300.

Moore, L., & Clark, D. E. (2008). The value of trauma registries. *Injury, 39*(6), 686–695.

Navsaria, P. H., Nicol, A. J., Krige, J. E., & Edu, S. (2009). Selective nonoperative management of liver gunshot injuries. *Annals of Surgery, 249*(4), 653–656. https://doi.org/10.1097/SLA.0b013e31819ed98d.

Newgard, C. D., Schmicker, R. H., Sopko, G., Andrusiek, D., Bialkowski, W., Minei, J. P., Brasel, K., Bulger, E., Fleischman, R. J., & Kerby, J. D. (2011). Trauma in the neighborhood: A geospatial analysis and assessment of social determinants of major injury in North America. *American Journal of Public Health, 101*(4), 669–677.

Nicol, A., Knowlton, L. M., Schuurman, N., Matzopoulos, R., Zargaran, E., Cinnamon, J., Fawcett, V., Taulu, T., & Hameed, S. M. (2014). Trauma surveillance in Cape Town, South Africa: An analysis of 9236 consecutive trauma center admissions. *JAMA Surgery, 149*(6), 549–556.

Norman, R., Matzopoulos, R., Groenewald, P., & Bradshaw, D. (2007). The high burden of injuries in South Africa. *Bulletin of the World Health Organization, 85*, 695–702.

Nwomeh, B. C., Lowell, W., Kable, R., Haley, K., & Ameh, E. A. (2006). History and development of trauma registry: Lessons from developed to developing countries. *World Journal of Emergency Surgery, 1*(1), 32.

O'Reilly, G. M., Cameron, P. A., & Joshipura, M. (2012). Global trauma registry mapping: A scoping review. *Injury, 43*(7), 1148–1153.

O'Reilly, G. M., Joshipura, M., Cameron, P. A., & Gruen, R. (2013). Trauma registries in developing countries: A review of the published experience. *Injury, 44*(6), 713–721.

O'Reilly, G. M., Gabbe, B., Braaf, S., & Cameron, P. A. (2016). An interview of trauma registry custodians to determine lessons learnt. *Injury, 47*(1), 116–124.

Owen, J. L., Bolenbaucher, R. M., & Moore, M. L. (1999). Trauma registry databases: A comparison of data abstraction, interpretation, and entry at two level I trauma centers. *Journal of Trauma and Acute Care Surgery, 46*(6), 1100–1104.

Owor, G., & Kobusingye, O. C. (2001). Trauma registries as a tool for improved clinical assessment of trauma patients in an urban African hospital. *East and Central African Journal of Surgery, 6*(1), 57–63.

Parkinson, F., Kent, S., Aldous, C., Oosthuizen, G., & Clarke, D. (2013). Road traffic crashes in South Africa: The burden of injury to a regional trauma centre. *South African Medical Journal, 103*(11), 850–852. https://doi.org/10.7196/SAMJ.6914.

Parreira, J. G., Campos, T. D. E., Perlingeiro, J. A. G., Soldá, S. C., Assef, J. C., Gonçalves, A. C., Zuffo, B., Floriano, C. G., Oliveira, E. H. D. E., & Oliveira, R. V. R. D. E. (2015). Implementation of the trauma registry as a tool for quality improvement in trauma care in a Brazilian hospital: The first 12 months. *Revista do Colégio Brasileiro de Cirurgiões, 42*(4), 265–272.

Pino Sánchez, F. I., Ballesteros Sanz, M. A., Cordero Lorenzana, L., & Guerrero López, F. (2015). Quality of trauma care and trauma registries. *Medicina Intensiva (English Edition), 39*(2), 114–123. https://doi.org/10.1016/j.medine.2015.02.002.

Racioppi, F., Eriksson, L., Tingvall, C., & Villaveces, A. (2004). *Preventing road traffic injury: A public health perspective for Europe* (p. 100). Copenhagen: World Health Organization.

Rainer, T. H., Cheung, N. K., Yeung, J. H. H., & Graham, C. A. (2007). Do trauma teams make a difference?. A single centre registry study. *Resuscitation, 73*(3), 374–381. https://doi.org/10.1016/j.resuscitation.2006.10.011.

Rosenkrantz, L., Shuurman, N., & Hameed, M. (2019). Trauma registry implementation and operation in low and middle income countries: A scoping review. *Global Public Health, 14*, 1884–1897.

Rosenkrantz, L., Schuurman, N., Arenas, C., Nicol, A., & Hameed, M. S. (2020). Maximizing the potential of trauma registries in low-income and middle-income countries. Trauma Surgery & Acute Care Open, *5*(1), e000469.

Rutledge, R. (1995). The goals, development, and use of trauma registries and trauma data sources in decision making in injury. *Surgical Clinics of North America, 75*(2), 305–326.

Schuurman, N., Hameed, S. M., Fiedler, R., Bell, N., & Simons, R. K. (2008). The spatial epidemiology of trauma: The potential of geographic information science to organize data and reveal patterns of injury and services. *Canadian Journal of Surgery, 51*(5), 389.

Schuurman, N., Cinnamon, J., Matzopoulos, R., Fawcett, V., Nicol, A., & Hameed, S. M. (2011). Collecting injury surveillance data in low- and middle-income countries: The Cape Town trauma registry pilot. *Global Public Health, 6*(8), 874–889. https://doi.org/10.1080/17441692.2010.516268.

Scott, D., Harrison, J., Purdie, D., Bain, C., Najman, J., Nixon, J., Spinks, A. B., & McClure, R. J. (2006). The properties of the International Classification of the External Cause of Injury when used as an instrument for injury prevention research. *Injury Prevention, 12*(4), 253–257.

Toroyan, T. (2009). Global status report on road safety. *World Health Organization, 15*(4), 286–286. https://doi.org/10.1136/ip.2009.023697.

Walker, B. B., Schuurman, N., & Hameed, S. M. (2014). A GIS-based spatiotemporal analysis of violent trauma hotspots in Vancouver, Canada: Identification, contextualisation and intervention. *BMJ Open, 4*(2), 1–7. https://doi.org/10.1136/bmjopen-2013-003642.

Weeks, S. R., Juillard, C. J., Monono, M. E., Etoundi, G. A., Ngamby, M. K., Hyder, A. A., & Stevens, K. A. (2014). Is the Kampala trauma score an effective predictor of mortality in low-resource settings? A comparison of multiple trauma severity scores. *World Journal of Surgery, 38*(8), 1905–1911. https://doi.org/10.1007/s00268-014-2496-0.

World Health Organization (WHO). (2014a). *Injuries and violence, the facts 2014.*

World Health Organization (WHO). (2014b). *International classification of diseases (ICD).* Retrieved from http://www.Who.Int/classifications/icd/en/(2011).

Zargaran, E., Schuurman, N., Nicol, A. J., Matzopoulos, R., Cinnamon, J., Taulu, T., Ricker, B., Garbutt Brown, D. R., Navsaria, P., & Hameed, S. M. (2014). The electronic trauma health record: Design and usability of a novel tablet-based tool for trauma care and injury surveillance in low resource settings. *Journal of the American College of Surgeons, 218*(1), 41–50. https://doi.org/10.1016/j.jamcollsurg.2013.10.001.

Zargaran, E., Spence, R., Adolph, L., Nicol, A., Schuurman, N., Navsaria, P., Ramsey, D., & Hameed, S. M. (2018). Association between real-time electronic injury surveillance applications and clinical documentation and data acquisition in a south African trauma center. *JAMA Surgery, 153*(5), 1–8. https://doi.org/10.1001/jamasurg.2018.0087.

Zavalaa, D. E., Bokongo, S., John, I. A., Senoga, I. M., Mtonga, R. E., Mohammed, A. Z., Anjango, W. O., & Olupot-Olupot, P. (2008). Implementing a hospital based injury surveillance system in Africa: Lessons learned. *Medicine, Conflict, and Survival, 24*(4), 260–272. https://doi.org/10.1080/13623690802373884.

Zehtabchi, S., Nishijima, D. K., McKay, M. P., & Clay Mann, N. (2011). Trauma registries: History, logistics, limitations, and contributions to emergency medicine research. *Academic Emergency Medicine, 18*(6), 637–643. https://doi.org/10.1111/j.1553-2712.2011.01083.x.

Chapter 12
Geography of Alcohol Exposure: Policy and Programme Implications for Cape Town, South Africa

Kim Bloch, Chris Berens, and Richard Matzopoulos

Abstract South Africa has some of the highest alcohol consumption levels worldwide. It is a major risk factor for health-related harms, which are amplified through the intersection with poverty and informality of settlements in areas such as Khayelitsha, Cape Town. Access to alcohol in low-income townships is facilitated by structural, supply-side factors including low prices of alcohol, high density of liquor outlets, poor regulation of sales and overt advertising. As such, alcohol consumption and problem drinking are higher in these low-income, marginalised areas. This chapter describes (1) existing literature on alcohol and alcohol-related harms in South Africa; (2) alcohol policy in the Western Cape Province, South Africa; (3) geographical characteristics of alcohol outlet exposure of residents in Khayelitsha, Cape Town, South Africa based on a comprehensive alcohol outlet mapping study; and (4) practical applications of these data for alcohol-related harms reduction policy and programming. The findings of this work are potentially useful to policy makers, programme implementers and law enforcement agents to more effectively regulate alcohol availability and reduce alcohol-related harms through evidence-based, spatial interventions.

Keywords Violence · Violence prevention · Liquor outlet mapping · Alcohol-related harm

K. Bloch (✉)
University of Cape Town, School of Public Health, Cape Town, South Africa

C. Berens
Violence Prevention Through Urban Upgrading (VPUU), Cape Town, South Africa

R. Matzopoulos
University of Cape Town, School of Public Health, Cape Town, South Africa

Burden of Disease Research Unit, South African Medical Research Council, Cape Town, South Africa

P. T. Makanga (ed.), *Practicing Health Geography*, Global Perspectives on Health Geography, https://doi.org/10.1007/978-3-030-63471-1_12

Background

Alcohol Use and Alcohol-Related Harms in South Africa

According to a World Health Organization (WHO) global status report, South Africa has the highest per capita (population aged 15 years and older) alcohol consumption in Africa—27.1 L of pure alcohol consumed by South African drinkers per capita (WHO 2014). The alcohol industry in South Africa substantially contributes to South Africa's economy, generating revenue of R93.2 billion, which equates to 2.9% of the total GDP (Western Cape Government 2017). Evidence from higher income country contexts as well as South African data suggest that profits from alcohol sales are dependent on excessive drinking amongst the minority of drinkers rather than widespread, moderate drinking (Western Cape Government 2017). Only an estimated 30–40% of South Africans consume alcohol, driving alcohol sales (WHO 2014; Peltzer et al. 2011). A population-level study of 15,828 South Africans found that men (41.5%) drank more than women (17.1%), urban residents (33.4%) were more likely to drink than rural residents (18.3%) and 9% reported risky, hazardous or harmful drinking (17% of men and 2.9% of women) (Peltzer et al. 2011).

Yet, alcohol is a major risk factor for health-related harms in South Africa, accounting for an estimated 7.1% of all deaths and 7% of total Deaths and disability-adjusted life years (DALYS) (Schneider et al. 2007). These alcohol-related harms greatly overshadow the economic benefits from the alcohol industry, leading to an annual net loss of R165–236 billion or 7–10% of the GDP (Western Cape Government 2017). The harmful effects of alcohol, such as crime and interpersonal violence, are amplified through the intersection with poverty, informality of settlements and cultural norms (Makanga et al. 2017). Duncan et al. (2011) describes a circular causality in the relationship between alcohol abuse and poverty; the authors attribute rising poverty levels to increased alcohol abuse and also the implications of alcohol abuse to low household income (Duncan et al. 2011).

Alcohol and Associated Disease Burden in Cape Town, South Africa

Khayelitsha, one of Cape Town, South Africa's poorest areas, is home to nearly 400,000 residents—just over 10% of the entire Cape Town population (City of Cape Town 2013). It is located approximately 35 km outside of the city centre and is covered by three South African Police Service (SAPS) precincts—Harare, Khayelitsha (Site B) and Lingelethu West, which span 38.7 km^2 (Fig. 12.1). The community experiences prevailing social, cultural, economic, institutional and spatial marginalisation. The majority of residents live in informal dwellings (54%), 64% have lower than Grade 12 education and 38% are unemployed (City of Cape Town 2013). In comparison to other Cape Town sub-districts, Khayelitsha has the

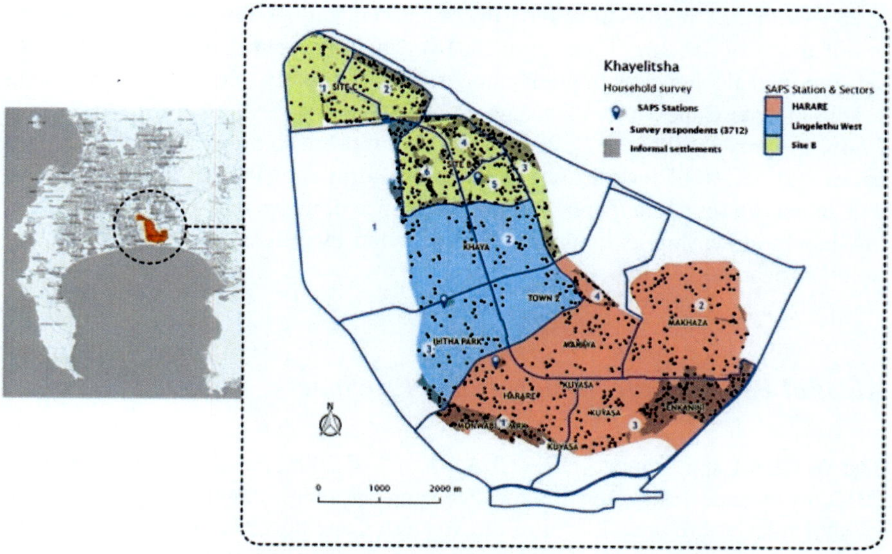

Fig. 12.1 Khayelitsha township situated within greater Cape Town

highest levels of mortality attributed to non-communicable diseases, HIV/AIDS, homicide, tuberculosis and road traffic injuries (Groenewald et al. 2010). Mortality rates related to homicide in Khayelitsha are double that of Cape Town and HIV/AIDS-related deaths are three-fold that of Cape Town (Groenewald et al. 2010).

Access to alcohol and harmful drinking patterns are facilitated by structural, supply-side factors including low prices of alcohol, high density of outlets, poor regulation and enforcement of sales and overt advertising (Western Cape Government 2017), which are exacerbated in low-income settings such as Khayelitsha relative to middle- and upper-income residential areas. According to the Khayelitsha Commission, there were 1400 unlicensed liquor outlets (locally known as shebeens) and 35 licensed shebeens in Khayelitsha reported in 2014 (Khayelitsha Commission 2014), a rate of 1 outlet for every 285 people. It is estimated that 70–80% of South African Brewery (SAB) products are consumed in the informal, unlicensed, predominantly unregulated market (Webster et al. 2008). The high density and easy access of alcohol outlets, particularly shebeens, have been reported as concerning to residents of Khayelitsha, particularly those outlets close in proximity to schools, which may potentially entice underage drinking (Matzopoulos et al. 2017; Puoane et al. 2013).

This chapter provides evidence that alcohol consumption and problem drinking is higher in the marginalized communities on the outskirts of Cape Town than in the general public. 'Problem drinking', as defined by Peltzer et al. (2011), is characterised by hazardous drinking (i.e. alcohol consumption that places individuals at risk for adverse health events) and harmful drinking (i.e. alcohol consumption that results in adverse events) (Peltzer et al. 2011). Peer et al. (2014) cited one's inability to cope with the stressors of poverty in conjunction with poor mental health as drivers of problem drinking (Peer et al. 2014). They reported that the prevalence of self-reported alcohol consumption amongst 1099 randomly selected residents living in low-income Cape Town townships in 2008/9 was 46.9% (68.5% in males and 27.4% in females) (Peer et al. 2014). Amongst those who drank alcohol, 73.4% of males and 67.1% of females reported problem drinking (Peer et al. 2014). These data indicate a four-fold increase in the prevalence of males who problem drink in a low-income township as compared data reported by Peltzer et al. on the general South African population.

Alcohol Policy in the Western Cape Province

The Western Cape Liquor Act (WCLA) (No. 4 of 2008) came into effect in April 2012, to provide legislation for the licensing of micro-manufacturing and retail alcohol sales at Provincial level in the Western Cape province. Implementation of the WCLA led to increasingly intense activity by law enforcement to restrict the number of illegal alcohol outlets, which was initially credited with a sudden reduction in violent crime. However, the act has been criticised due to poor planning, increased relaxation of regulatory controls, inability to provide adequate or appropriate law enforcement, lack of accountability of law enforcement, lack of collaboration between the necessary stakeholders and poor surveillance (Faull 2013; Myers 2015).

The WCLA was gradually corroded by liquor industry efforts including their political influence that saw a gradual relaxation of its most effective provisions around alcohol access. The Khayelitsha Commission, an independent Commission of Inquiry, noted regulatory challenges posed by unlicensed alcohol outlets, the potential implications of stricter law enforcement on traders' livelihoods and relationships between police and community members as further barriers to WCLA implementation (Khayelitsha Commission 2014). The inevitable increase in alcohol-related harms has seen the Western Cape Government revisit their alcohol policy, with the recent drafting of the Green paper and White paper.

Prevailing alcohol-related harms culminated in the Western Cape Provincial Cabinet committing to develop a more conducive alcohol-related harms reduction policy in 2015 that took a more comprehensive harm-reduction approach founded on the WHO global strategy to reduce the harmful use of alcohol using a ten-point plan (World Health Organization 2010). The ten-point plan targets ten focal areas including leadership, health services responses, community action, drink-driving policies, alcohol availability, marketing, pricing, reducing the negative consequences of drinking, reducing the public health impact of illicit alcohol and monitoring (World Health Organization 2010).

The Western Cape Government policy intended to address the current legislation's shortcomings by transitioning to a 'whole of society' framework, which places a stronger emphasis on the primary and secondary impacts of alcohol-related harms on society and calls for integrated action between all levels of stakeholders (Western Cape Government 2017). Policy development was guided by a diverse public-sector working group comprised of representatives from government, SAPS, Western Cape Liquor Authority and academia (Western Cape Government 2017). The Green Paper for the 'Western Cape Alcohol-Related Harms Reduction Policy' was published in September 2016 for public comment and a regulatory impact assessment was conducted—both of which informed finalisation of the White Paper published in August 2017 (Western Cape Government 2017). The White Paper consists of nine target policy areas relating to pricing, licensing, law enforcement, the road environment, health and social services, community participation, education and awareness, monitoring and evaluation, and institutional involvement in policy roll-out (Western Cape Government 2017). At the time of writing this chapter, high-level Alcohol Harms Reduction Liquor Bill drafting workshops weer underway as part of the next steps towards legislating the Alcohol Harms Reduction policy.

The availability of alcohol is a key determinant of adverse health outcomes, including injuries and the World Health Organisation identified alcohol availability as one of its key target areas in its Global Alcohol Strategy. Regulating the number and location of on-premise and off-premise alcohol outlets and regulating days and hours of retail sales are among the key strategies to reduce access. The Western Cape Alcohol-Related Harms Reduction Policy contains provisions to reduce access according to both strategies. A systematic review of studies published from 2000 to 2008 on the hours and days of sale and density of alcohol outlets found that the restrictions impacted on three main outcomes: overall alcohol consumption, drinking patterns and damage from alcohol. Several published literature reviews have considered the function of geography on alcohol consumption by examining the effects of alcohol outlet densities (Livingston 2011) or proximity to outlets (Hay et al. 2009; Ngamini Ngui et al. 2015; Murray and Roncek 2008). However, there is seldom strong empirical basis for the limits that are set, other than that they were more or less restrictive than were previously legislated within those same jurisdictions. In addition, most literature is from the global north where the alcohol trade environment is fundamentally different, which further complicates the wholesale adoption of standards from another setting.

This chapter describes (1) existing literature on alcohol and alcohol-related harms in South Africa; (2) alcohol policy in the Western Cape Province, South Africa; (3) geographical characteristics of alcohol outlets exposure of residents in a low-income township in Cape Town, South Africa; and (4) practical applications of these data for alcohol-related harms reduction policy and programming. This study offers the most comprehensive geographic and descriptive data available to date on the extent of the alcohol trade in one of the largest low-income townships in Cape Town.

Modelling Exposure to Liquor Outlets: Cape Town, South Africa

Mapping Location and Capacity of Liquor Outlets

Comprehensive alcohol outlet mapping was conducted in two phases: (1) April to May 2015 and (2) August 2016. Phase one covered 70% of Khayelitsha and phase two covered the remaining 30%. Trained fieldworkers recruited from the Khayelitsha community systematically conducted fieldwork street-by-street to identify, geolocate and gather information about all accessible alcohol outlets in Khayelitsha. These data included, inter alia, type of outlet per visual cues, size of outlet, years in operation, operating hours, characteristics of patrons and information about licensing. The fieldworkers used aerial maps for navigation and a secure, open-source, GPS enabled, mobile data collection software called Open Data Kit (ODK) to collect data. A detailed description of the methodology and detailed analysis on outlet characteristics has been previously published by Matzopoulos et al. (2017, 2019).

Defining Alcohol Outlet Exposure

While alcohol outlet density (AOD) provides a ratio of alcohol outlets to population for a given spatial extent it does not take in to account the effect on the resident population of the outlets in adjacent areas. The effect of an outlet is not limited by administrative jurisdiction but more likely by proximity. Given that this study has access to the GPS point locations of the outlets across the study area it was possible to model an alcohol outlet exposure (AOE) based on proximity.

Modelling an Alcohol Outlet Exposure Index

Included in the study was a measure of size or capacity. Of all outlets, 63.2% were considered 'small' (capacity of 0–39 patrons) and 22.2% were 'large' (capacity of ≥40 patrons). Fieldworkers were unable to determine the capacity of 14.6% of outlets. To account for varying capacity of outlets, their effects were weighted by size; takeaways were allocated a weighted score of 0.5, small outlets as 1, large outlets as 1.5 and outlets with unknown capacity as 1.

Using the GPS point locations of the outlets, a heat map of exposure to alcohol outlets was generated giving overage of the entire study area. The effect of each outlet was modelled using a tri-weight kernel with a radius of 500 m. This model applies a non-linear score from one at source to zero at 500 m distance from the outlet. The non-linear approach models the indirect consequences of alcohol outlets being situated in residential spaces, including noise pollution, public drunkenness and damage to property. A 500 m radius was chosen to ensure the entire study area was covered with an indexed score greater than zero, allowing for the capacity to measure variation in exposure.

Results: Exposure to Alcohol

Characteristics of Alcohol Outlets in Khayelitsha

A total of 1045 alcohol outlets were identified in Khayelitsha during the study period[1]—958 during phase one and an additional 87 during phase two[1]. Nearly all (92%) alcohol outlets in Khayelitsha operate within residential areas. Of those, 606 were located in predominantly formal residential areas and 356 in informal settlements.

Fieldworkers visually classified alcohol outlets into four defined types:

- Shebeen: unlicensed outlet for on-site consumption.
- Take-away: unlicensed outlet with no on-site consumption.
- Umqombothi: unlicensed outlet that sells traditional beer for on-site consumption.
- Tavern: licensed outlet for on-site consumption.

The vast majority of outlets were shebeens (71.8%), and of those, 89.3% were located in Site B precinct. The second most common type was taverns (16.3%). Less commonly, 9% of identified outlets were takeaways and 2.9% umqombothis.

In Khayelitsha, there was an overall ratio of one outlet per every 375 Khayelitsha residents (Fig. 12.2). Alcohol outlet density and therefore accessibility varied by SAPS station and there were distinct hotspots: Site B precinct has the highest ratio of 1 outlet per every 334 residents, followed by 1:383.5 in Harare precinct and 1:484.5 in Lingelethu West precinct.

Alcohol Outlet Exposure of Areas in Khayelitsha

A heat map constituting an alcohol outlet exposure index was generated using Geographical Information Systems (GIS) by measuring the sum of the overlapping effects of each outlet on a 5-m grid across the study area (Fig. 12.3). The resulting model allows for interrogation of the alcohol exposure at any reference point (e.g. household) or area (e.g. police sector) within the study area.

[1] 1076 outlets were located; however, 31 outlets were excluded due to apparent data collection errors that could not be rectified.

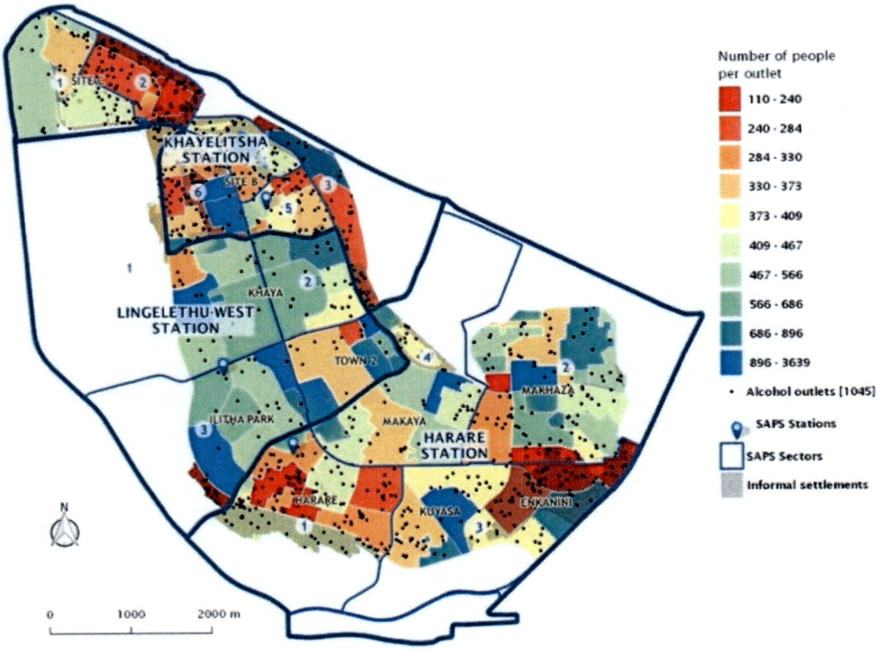

Fig. 12.2 Ratio of residents per outlet by SAPS geoblock

The median alcohol outlet exposure score in Khayelitsha was 6.53. Conceptually, this means that Khayelitsha residents at any given location or time were 'exposed' to the cumulative effects of an average of more than six 'small' alcohol outlets, within 500 m from their residence. When restricting the area of analysis to residential areas only, the median exposure score increased to 7.47, which is to be expected given that 92% of outlets are located in residential areas. In informal areas, the median exposure score was 9.54, which is 38% higher than formal areas (6.93).

Table 12.1 and Fig. 12.4 describe median exposure scores stratified by SAPS station and sector boundaries and corresponding ranking. Site B station had the highest median exposure score of 11.41 followed by Harare (median exposure score 5.25) and Lingelethu West (median exposure score of 2.6). Sectors within Site B had the top three highest median exposure scores (i.e. the highest outlet dense areas with the largest cumulative exposure effects): Sector 2 (median score 22.53), Sector 6 (median score 19.93) and Sector 4 (median score 17.5). Lingelethu West had the lowest scores: Sector 3 (median score 2.28), Sector 1 (median score 2.25), and Sector 4 (median score 2.22). Figure 12.5 provides a more granular perspective of median alcohol exposure scores by SAPS geoblock. Using this map, pockets of higher than average alcohol exposure within a particular SAPS station, such as in the south-western area of Harare, can be easily identified.

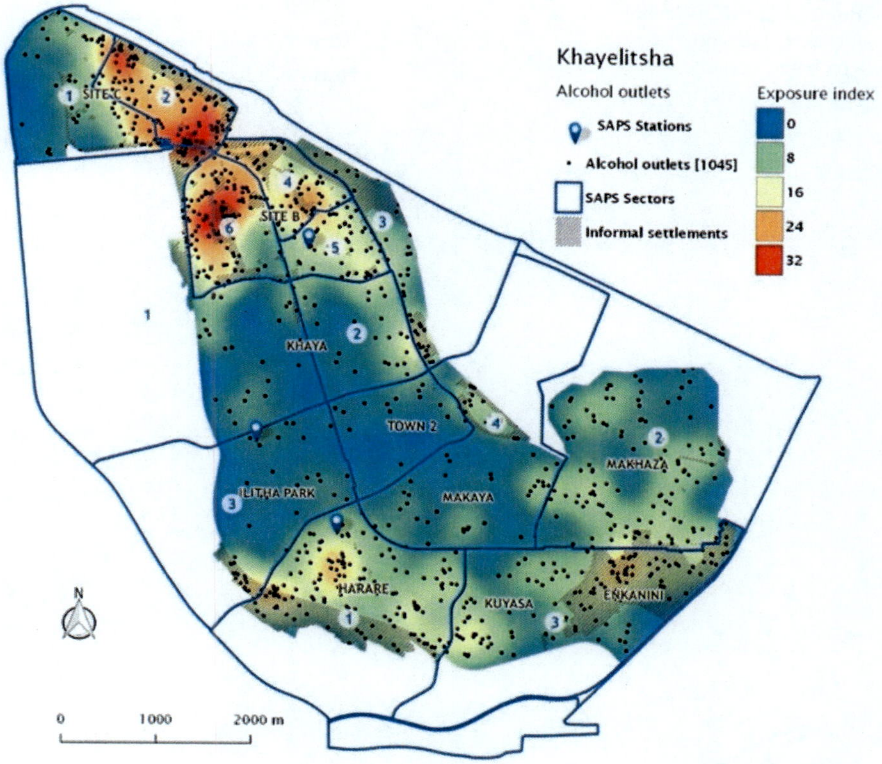

Fig. 12.3 Outlet exposure index heatmap for Khayelitsha

Applications and Recommendations Policy and Programming

To our knowledge, this is the most comprehensive study of its kind to map alcohol outlet density and define an alcohol outlet exposure index in a low-income township in Cape Town, South Africa. These data have the ability to uniquely illustrate and illuminate the extent of the alcohol trade in Khayelitsha and the cumulative effects of alcohol outlet exposure on Khayelitsha residents. This includes identification of geographic hotspots where there are clusters of alcohol outlets and high outlet exposure of residents. With this information, policy makers, programme implementers and law enforcement can more effectively regulate alcohol availability and alcohol-related harms through area-based interventions.

A powerful aspect of this dataset is its adaptability. Alcohol outlet attributes data are linked to GPS coordinates, which allow for spatial analyses using any geographic boundaries of interest from SAPS precincts down to street-level. ***It is recommended to interrogate these data across different spatial hierarchies, which will offer varying levels of granularity in identifying spatial patterns and hotspots.***

Table 12.1 Median alcohol outlet exposure scores by SAPS boundaries and corresponding ranking

Area	Median score	Ranking
Harare	**5.25**	**II**
1	8.74	5
2	4.62	10
3	6.15	6
4	3.52	11
Lingelethu west	**2.6**	**III**
1	2.25	13
2	5.12	8
3	2.28	12
4	2.22	14
Site B	**11.41**	**I**
1	4.92	9
2	22.53	1
3	5.66	7
4	17.5	3
5	13.95	4
6	19.93	2

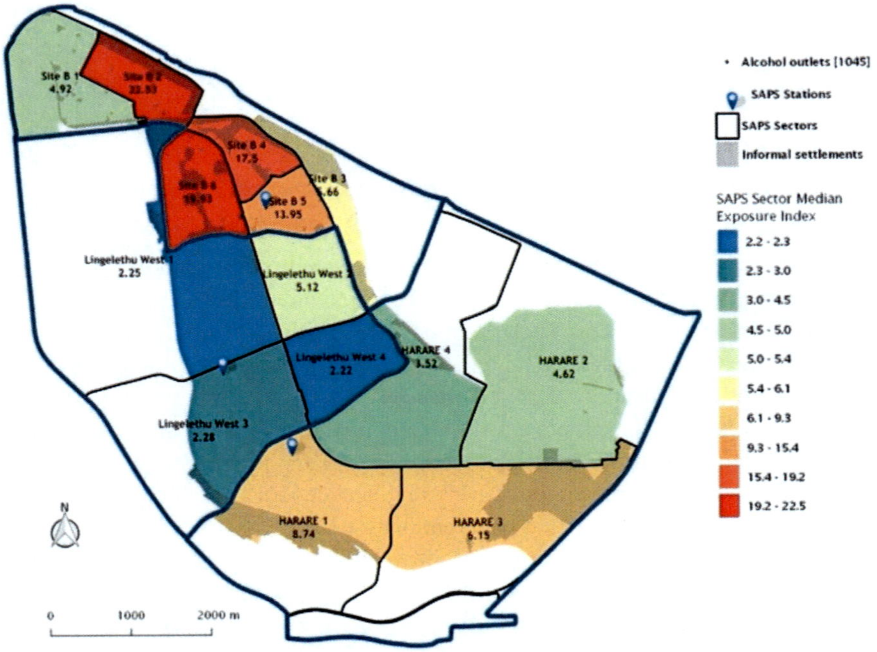

Fig. 12.4 Median alcohol outlet exposure scores by SAPS sector

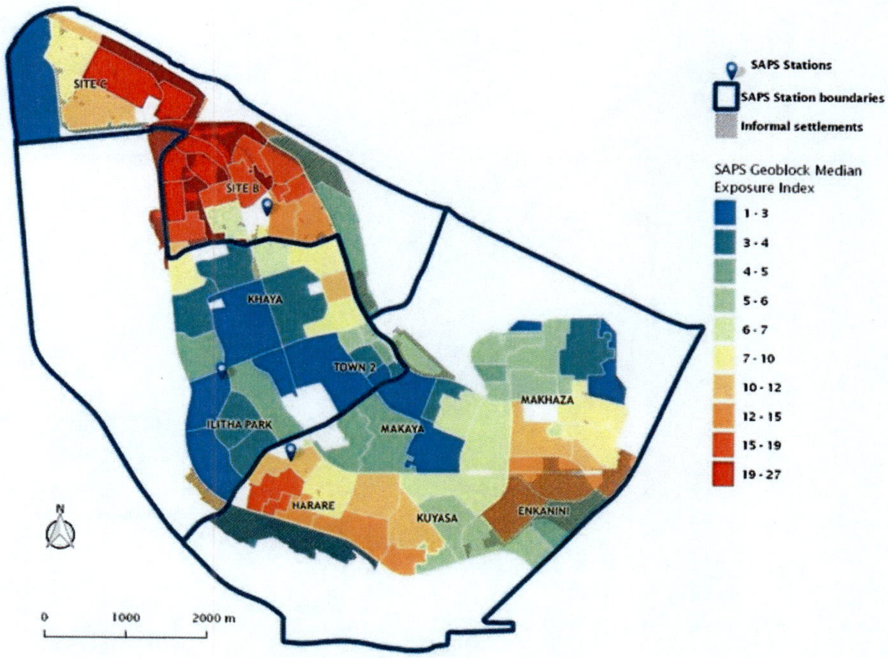

Fig. 12.5 Median alcohol outlet exposure scores by SAPS geoblock

Prior to this study, geographic areas in the Cape Town Metropole such as Lingelethu West, Khayelitsha were targeted by the Western Cape Government as pilot sites for an alcohol harms reduction programme. According to our data, Lingelethu west has the lowest alcohol outlet density compared to other SAPS stations across Khayelitsha. With access to these data, a more effective evidence-based approach could be taken to identify hotspots such as Site B, Khayelitsha (e.g. sectors 2, 6 and 4) for targeted interventions. Furthermore, evidence from Matzopoulos et al. suggests more violence and poorer mental health in Site B than Lingelethu West (Fig. 12.5) (Matzopoulos et al. 2017, 2019), which, per existing literature, is likely to be partially attributable to the high density of alcohol outlets in the area. ***It is recommended that an evidence-based approach is taken to identify hotspots with high alcohol outlet densities for targeted priority areas for alcohol-harms reduction strategies.***

To further investigate this association, alcohol outlet exposure data could be merged with health outcomes data linked to geolocations, such as violence data from a household survey with corresponding GPS coordinates. The proposed methodology for this analysis would include overlaying the alcohol outlet exposure heatmap with GPS points of households surveyed, querying the value of the exposure index at each household and linking the score to their household survey record (Fig. 12.6).

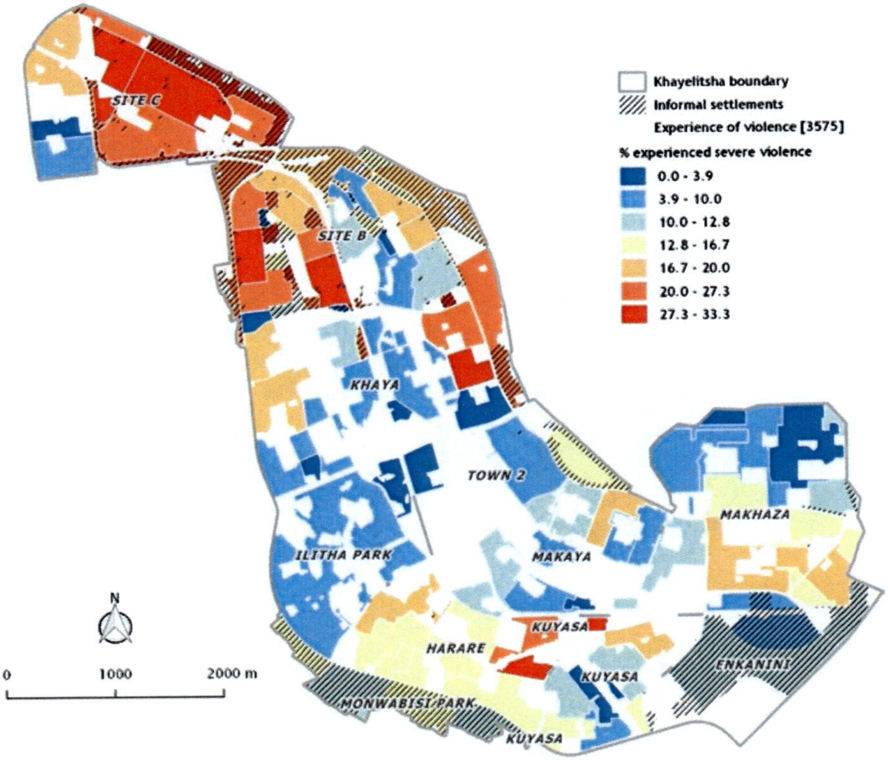

Fig. 12.6 Severe interpersonal violence map adapted from the SAIC chapter

This would enable one to investigate the relationship between alcohol outlet exposure and violence using spatial mapping or biostatistical techniques such as multiple regression modelling to adjust for confounders. ***It is recommended that more sophisticated analysis of outlet exposure and social/health outcomes be conducted to inform prevention interventions, intensive screening services and targeted health services in areas with high alcohol outlet clustering and alcohol-related harms.***

These data can also play a valuable role in determining evidence-based targets for alcohol outlet density as a way to regulate alcohol outlet licensing and curtail alcohol-related harms. The proposed targets could be area-based and consulted when considering alcohol outlet licensing applications or developing the most effective strategies for outlet closures in highly clustered areas. A proposed methodology to determine an appropriate cut-off would be identify an appropriate outcome indicator such as number of violent crimes, interrogate the spatial distribution of these crimes and determine an appropriate cut-off that allows for a limited number of outlets whilst still maintaining below average crime rates. ***It is recommended that more sophisticated analysis of outlet exposure and social/health outcomes be***

conducted to inform geographic alcohol outlet density targets to work towards as a strategy to reduce alcohol-related harms.

Lastly, these data along with previously published data (Matzopoulos et al. 2017, 2019) can be used as a comprehensive baseline for monitoring and evaluating the impact of alcohol harms reduction policy and programming in Khayelitsha. ***It is recommended that this study's methodology be replicated in future studies and these data be considered a baseline for follow-up as part of assessing impact of policy for reducing access to alcohol.***

Facilitators to Drive Evidence-Based Policy and Programming

Whilst it is evident that these data provide valuable insights to support evidence-based policy and programming, key facilitators should be put in place to ensure the uptake of data-driven policy and programming:

- Researchers and policy makers should begin their collaboration during the early stages of developing research or policy agendas and outline clear linkages from the outset. In the case of a research study, this would require defining relevant research questions in consultation with stakeholders that result in more directed policy recommendations.
- Researchers should devise a dissemination strategy to engage with stakeholders around key findings using language that is accessible beyond the realm of academia. For example, this could include translation of key findings into tangible infographics. As part of the dissemination strategy, there should also be facilitated discussions around key policy and programming implications.
- There needs to be political will from key stakeholders to adopt an evidence-based approach. Gaining buy-in requires academic partnerships and education about the importance of an evidence-based approach founded on strong study design. For example, this would be pertinent in deconstructing preconceived notions of 'high-risk areas' based on anecdotal rather than empirical evidence.
- Sufficient time and funding through a research-policy partnership should be prioritised in order to effectively adopt an evidence-based approach.

Conclusions

South Africa has the highest per capita alcohol consumption (of alcohol drinkers) in Africa. It is well established that alcohol is a major risk factor for health-related harms and accounts for a substantial proportion of deaths and DALYS in South Africa. The harmful effects of alcohol are amplified through the intersection of poverty, informality of settlements and cultural norms, leaving low-income townships such as Khayelitsha highly susceptible to alcohol harms.

Prior to this study, there was a gap in knowledge regarding the extent of the alcohol trade in this area. This study provided valuable insights into the geographic distribution of alcohol outlets, geographic hotspots with high alcohol outlet densities and the cumulative effects of alcohol outlet exposure on Khayelitsha residents. In Khayelitsha, there was an overall ratio of one outlet per every 375 residents and more than two-thirds of outlets were informal and likely to be unlicensed, resulting in unregulated sales and potentially unsafe activities. At any given location or time, Khayelitsha residents were 'exposed' to the cumulative effects of an average of more than six 'small' alcohol outlets, within 500 m from their residence.

These findings have the power to underpin evidence-based policy and programming for alcohol harms reduction strategies for high-risk areas. With these data, geographic hotspots with high alcohol outlet densities can be identified and the effects of alcohol outlet density on social/health outcomes can be investigated geospatially. These data may also act as a comprehensive baseline for comparison over time as part of monitoring activities. However, essential to the translation of these data for effective policy/programming includes multi-sectoral collaboration, political will, appropriate funding and resources and a dissemination strategy that reaches key stakeholders.

Acknowledgements We would like to acknowledge the UK Government's Department for International Development (DFID), International Development Research Centre (IDRC), Canada (Grant numbers 107329-001 and 107202-002) and Department of Community Safety, Western Cape Government for their financial support. Thank you to all our collaborators and participants who are involved in the research.

References

Commission of Inquiry into Allegations of Police Inefficiency and a Breakdown in Relations between SAPS and the Community in Khayelitsha. (2014). TOWARDS A SAFER KHAYELITSHA: Report of the Commission of Inquiry into Allegations of Police Inefficiency and a Breakdown in Relations between SAPS and the Community of Khayelitsha. 1-579. ISBN: 978-1-920677-71-8.

Duncan, M., Swartz, L., & Kathard, H. (2011). The burden of psychiatric disability on chronically poor households: Part 1 (costs). *South African Journal of Occupational Therapy, 41*, 55–63.

Faull, A. (2013). Policing taverns and shebeens: Observation, experience and discourse. *South African Crime Quarterly*. https://doi.org/10.17159/2413-3108/2013/v0i46a805.

Groenewald, P., et al. (2010). Local-level mortality surveillance in resource-limited settings: A case study of Cape Town highlights disparities in health. *Bulletin of the World Health Organization, 88*, 444–451.

Hay, G. C., Whigham, P. A., Kypri, K., & Langley, J. D. (2009). Neighbourhood deprivation and access to alcohol outlets: A national study. *Health & Place, 15*(4), 1086–1093.

Livingston, M. (2011). Alcohol outlet density and harm: Comparing the impacts on violence and chronic harms. *Drug and Alcohol Review, 30*(5), 515–523.

Makanga, P. T., Schuurman, N., & Randall, E. (2017). Community perceptions of risk factors for interpersonal violence in townships in Cape Town, South Africa: A focus group study. *Global Public Health, 12*, 1254–1268.

Matzopoulos, R., Berens, C., Bowman, B., Bloch, K., Langa, M., Kotze, E., Lloyd, S., Makanga, PT., Mureithi, L., Schuurman, N. (2017). A mixed methods study of the nature and extent of the alcohol trade in Khayelitsha - and community perceptions. *Cape Town: Department of Community Safety.* ISBN: 978-1-928340-23-2.

Matzopoulos, R., Bloch, K., Lloyd, S., Berens, C., Myers, J., Thompson, ML. (2019). Urban upgrading linked to positive social outcomes in Cape Town, South Africa. *Reducing Urban Violence in the Global South: Towards Safe and Inclusive Cities.* 69-89. ISBN: 9781351254649.

Murray, R. K., & Roncek, D. W. (2008). Measuring diffusion of assaults around bars through radius and adjacency techniques. *Criminal Justice Review, 33*(2), 199–220.

Myers, J. (2015). Rapid review of the problem of alcohol harm reduction in the Western Cape province: Evidence-based effective and cost-effective interventions of global and local relevance, barriers to uptake, and recommendations for policy development and implementation. *Cape Town: University of Cape Town.*

Ngamini Ngui, A., Apparicio, P., Philibert, M., & Fleury, M.-J. (2015). Neighborhood characteristics associated with the availability of alcohol outlets in Quebec, Canada. *Journal of Addiction, 2015*, 1.

Peer, N., Lombard, C., Steyn, K., & Levitt, N. (2014). Rising alcohol consumption and a high prevalence of problem drinking in black men and women in Cape Town: The CRIBSA study. *Journal of Epidemiology and Community Health, 68*, 446–452.

Peltzer, K., Davids, A., & Njuho, P. (2011). Alcohol use and problem drinking in South Africa: Findings from a national population-based survey. *The African Journal of Psychiatry, 14*, 30–37.

Puoane, T. R., Tsolekile, L., & Sanders, D. (2013). A case study of community-level intervention for non-communicable diseases in Khayelitsha, Cape Town. *Brighton: Institute of Development Studies (IDS).*

Schneider, M., et al. (2007). Estimating the burden of disease attributable to alcohol use in South Africa in 2000. *South African Medical Journal (Suid-Afrikaanse tydskrif vir geneeskund), 97*, 664–672.

Strategic Development Information and GIS Department (SDI&GIS), City of Cape Town. (2013). *City of Cape Town—2011 census suburb Khayelitsha.*

Webster, E., Benya, A., Dilata, X., Joynt, C., Ngoepe, K., & Tsoeu, M. (2008). Making the visible invisible: Confronting South Africa's decent work deficit, research commissioned by the Department of Labour.

Western Cape Government. (2017). Western Cape alcohol-related harms reduction policy White Paper.

World Health Organization. (2010). Global strategy to reduce the harmful use of alcohol. *Geneva: World Health Organization.*

World Health Organization. (2014). Global status report on alcohol and health 2014. *Geneva: World Health Organization.*

Chapter 13
Geography, Climate Change and Health Adaptation Planning in Uganda

Didacus B. Namanya, Lea Berrang-Ford, Sherilee L. Harper, James Ford, Everd M. Bikaitwoha, Shuaib Lwasa, Carlee J. Wright, and Christopher Kazaana

Abstract Public health interventions in many African settings have traditionally been driven by data on disease burden. Geography, especially in the context of climate change, is being incorporated into the process of health planning to enunciate the spatial and place-specific nature of health and disease. This chapter will describe a case of how geography is used to plan for population level health interventions in regions of Uganda. In particular, the authors emphasize the intersection between climate change and public health planning by examining Uganda's disease burden in relation to climate change and geography. We also explore climate change health vulnerabilities, climate variability and climate change health adaptation planning. Relevant policy frameworks, their gaps, and how geographic thought can guide targeted action to lessen the potential negative effects of the changing climate on health in Uganda are also described.

Keywords Uganda · Climate change · Geography · Health adaptation planning

---

D. B. Namanya (✉)
Ministry of Health, Kampala, Uganda

Uganda Martyrs University, Kampala, Uganda

L. Berrang-Ford · J. Ford
University of Leeds, Priestly International Centre for Climate, Leeds, UK

S. L. Harper
University of Alberta, School of Public Health, Edmonton, AB, Canada

E. M. Bikaitwoha
Virtual University of Uganda, Kampala, Uganda

S. Lwasa
Makerere University, Kampala, Uganda

C. J. Wright
Population Medicine, University of Guelph, Guelph, ON, Canada

C. Kazaana
Makerere University Business School, Kampala, Uganda

P. T. Makanga (ed.), *Practicing Health Geography*, Global Perspectives on Health Geography, https://doi.org/10.1007/978-3-030-63471-1_13

Background

Climate change is a major challenge for humanity and is likely to disrupt human and natural systems, including health systems (Wang and Horton 2015). It is generally recognized that climate change impacts will be felt more in low and middle-income countries like Uganda, which are more vulnerable because the physical, economic, and social infrastructure are generally less prepared to respond and adapt (Wang and Horton 2015). In the last 10 years, the health sector in Uganda has undergone many reforms in the areas of governance, budgeting, service delivery, and monitoring. These reforms provide evidence of an institutionalized response to a changing landscape in national policy, economic status, demography, lifestyles, and disease burden (Ministry of Health, MOH 2016a). There are present opportunities to incorporate climate change adaptation considerations into the health sector.

Uganda is a land-locked country, lying astride the equator in eastern Africa. One third of the country is covered by fresh water and wetlands (Fig. 13.1). It is mainly a plateau, with the Rwenzori (5110 m above sea level) and Elgon (4321 m above sea level) mountains in the west and east, respectively. It has a tropical climate, with mean annual temperatures ranging between 16 °C in the southwest to 31 °C in the northeast. The western central, and eastern regions of Uganda experience two peaks in rainfall: one from March to May, and another from September to December. The northern parts of the country experience less rainfall and have one rainy season per year (June to August).

Uganda's evolving health system is skewed towards curative rather than preventive approaches, and faced with substantial health system and service delivery challenges. Amidst these public health challenges, there have been observed changes and variability in temperature, precipitation, and frequency and intensity of extreme weather events, resulting in diverse impacts on human health and wellbeing in

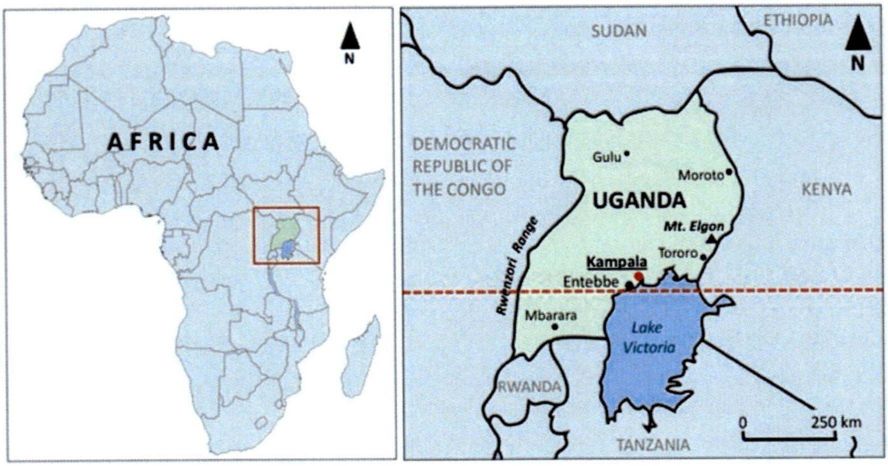

Fig. 13.1 Location of Uganda and key geographic features

Uganda (MWE 2015; WHO 2015). In this chapter, we advance the notion that Uganda's diverse geography has important implications for the health system, particularly disease patterns, climate variability, and change, as well as associated vulnerabilities and public health strategic planning and adaptation.

The Geography and Burden of Climate-Sensitive Diseases in Uganda

Climate-sensitive preventable diseases dominate Uganda's disease burden. According to the Ministry of Finance Planning and Economic development—MFPED (2015), malaria accounted for 37% of the overall disease burden, followed by non-pneumonia coughs/colds (22.3%), intestinal worms (5.4%), skin diseases (3.2%), acute diarrhoea (3.0%), and eye conditions (2.4%), among others (Fig. 13.2). Rising temperatures and more frequent extreme weather events like heavy rains, drought, wind storms, and heat and cold waves can cost lives, directly increase transmission and spread of diseases, and undermine the environmental determinants of health, including clean air, water, sufficient food, and secure shelter (MOH 2016b, p. 44).

The WHO (2015, p. 2) further indicates that existing literature shows many of the largest health concerns are strongly influenced by weather and climate conditions, which inevitably present risks for human health. These risks can be grouped into: (1) direct impacts, such as those arising from damages and illness due to increased

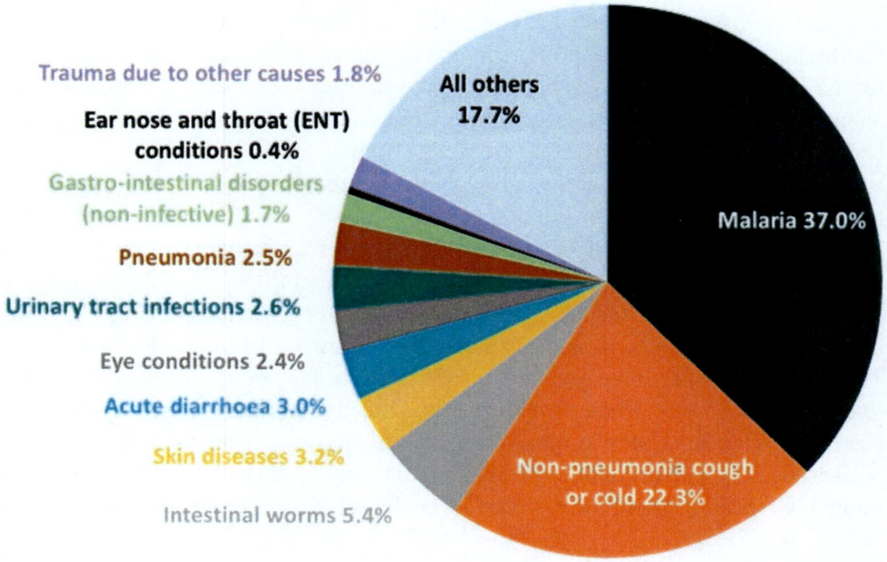

Fig. 13.2 Uganda's disease burden. (Source: MFPED 2015)

frequency and severity of extreme weather events; (2) environmental system-mediated impacts, such as rising air pollution and changing patterns of vector-, food- and water-borne diseases; and (3) socially mediated effects that occur via climate change impacts on social and human systems. These include health effects resulting from under nutrition, occupational heat stress, and mental illness, as well as potential increases in population displacement, slowing of economic growth, and poverty aggravation.

These impacts are also intricately linked with place and with the geography of the areas affected. Research from geography, epidemiology, and public health shows that where people live could significantly affect their health outcomes (Tunstall et al. 2004). The importance of 'place' to health status became increasingly clear in the last decades of the twentieth century. The characteristics of place include the social and economic environments, as well as the natural environment (e.g. air, water) and the built environment, which may include transportation, buildings, green spaces, roads, and other infrastructure (IOM 2001). In this chapter, we focus in particular on cholera and malaria, two of the most common and climate-sensitive diseases, to illustrate the relationship between climate, geography, and health in Uganda.

Cholera

Cholera is a preventable and treatable acute diarrhoeal disease caused by infection of the intestine with the bacterium *Vibrio cholerae*. Cholera is usually transmitted through consumption of water or food contaminated by faeces bearing the cholera bacterium. According to the MOH (2017), from the year 2011 to 2015, Uganda registered 9954 cholera cases. The majority of cases occurred in border districts. The highest number of cases were recorded during an outbreak in 2012, accounting for 6226 cases and 135 deaths. This outbreak was attributed to an *El Nino* phenomenon that is characterized by heavy rains and flooding. Similar findings had been previously reported in a separate study (Namanya 2009).

The districts historically affected by cholera outbreaks are geographically similar, typically bordering lakes or major wetlands; these geographical features present local populations with challenges in construction of pit latrines; a key sanitation facility. These districts are also prone to land slides and flooding which can trigger contamination of water, and thus, cholera outbreaks. A critical look at the 2011–2016 cholera map shows a geographic belt along the western border of Uganda (Fig. 13.3). These districts either share a boundary with a major lake or are in close proximity to one. Another visible cholera belt is geographically focused around Mt. Elgon. In Kasese district, for example, prolonged community-wide cholera outbreaks have been associated with drinking water contaminated by faecal matter and cross-border trading where some traders are careers of cholera (Kwesiga et al. 2018).

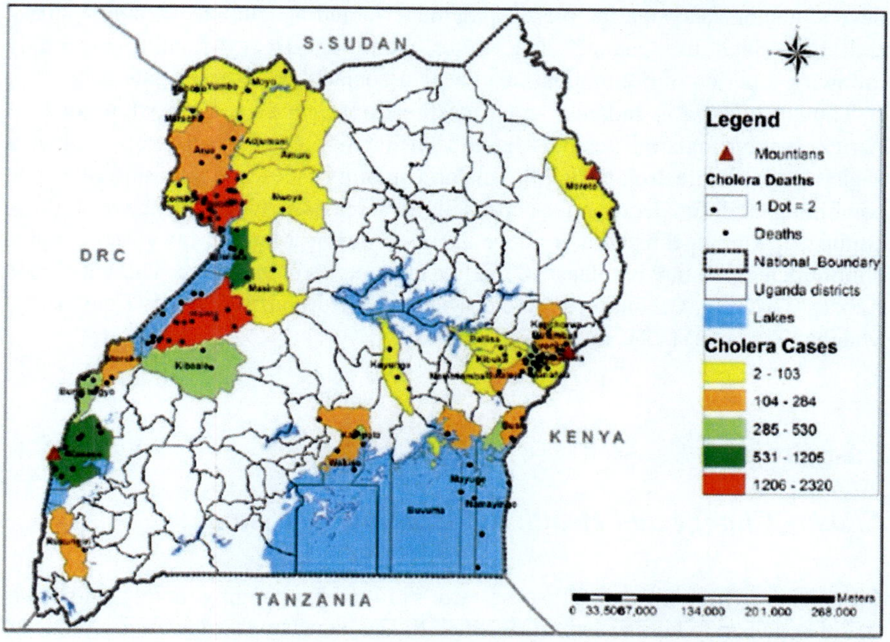

Fig. 13.3 Map of Uganda showing districts that reported cholera cases between 2011 and 2016. (Adopted from Ministry of Health 2017, p. 4)

Malaria

Malaria is one of the leading causes of ill health and death in Uganda, with approximately 16 million cases and over 10,500 deaths reported in 2013, accounting for 30–50% of outpatient visits and 15–20% of hospital admissions MOH (2014). The majority of climate-malaria research in Africa suggests that malaria transmission, especially epidemic outbreaks, is associated with increased rainfall in typically dry regions, and increased temperatures in high altitude and typically cool regions (Connor et al. 2006).

Unlike cholera, malaria is endemic in approximately 95% of Uganda, affecting over 90% of the population. The remaining 5% of the country consists of unstable and epidemic-prone transmission areas in the highlands of the south- and mid-west, along the eastern border with Rwanda, and the north-eastern border with South Sudan MOH (2014). Due to increasing temperature, malaria may become prevalent in areas that were previously free of malaria, especially highlands (EAC and USAID 2018). Barnett and Adger (2007) established that mosquitoes (*Anopheles*) are known to expand their range, coinciding with warming temperatures and increased precipitation. Risk of malaria transmission may therefore increase as pathogens change and survive longer, and as the migration pathways of carriers and vectors are altered. EAC and USAID (2018) also confirmed that malaria epidemics have been increasing in frequency in the highlands of eastern Africa over the past two decades.

Several factors account for these epidemics, including climate variability, drug resistance, land use change, and vector migration. Heavy rainfall, especially following a period of drought, has also been associated with these epidemics.

Though malaria is endemic and widespread in Uganda and the whole of East Africa, certain malarial hotspots have been identified, where geography plays a decisive role in infection. Using the malaria stability index and exposure map, and combining additional criteria, such as if the area is a highland above 1500-m altitude, if rainfall threshold is above 150 mm during March–May season, and if minimum temperature is at least 18 °C, five hotspots were identified. These included Kabale (Uganda), Gikonko (Rwanda), Muhanga (Burundi), Kericho (Kenya), and Muleba (Tanzania) (EAC and USAID 2018).

Case Studies

Climate Change and Health Vulnerability in Uganda

There is now strong evidence that the earth's climate is changing rapidly, and this is due in large part to human activities (IPCC 2007). Increasing temperatures, sea level rise, changes in precipitation patterns, and extreme events are expected to increase a number of health risks. Direct effects of climate change include heat waves, floods, and storms, and more suitable conditions for the transmission of important infectious diseases. Indirect effects include impacts on natural systems and socioeconomic sectors that ultimately underpin human health (WHO 2013). Vulnerability to climate change is the degree to which a system is susceptible to, or unable to cope with, the adverse effects of climate variability and change (IPCC 2007).

Climate change impacts have manifested and been documented in Uganda, including increased frequency and severity of extreme weather events, increased drought, changes in precipitation patterns, and shifts in seasons (Magrath 2008; Namanya 2009, Hepworth 2010; Hussein 2011; Hartter et al. 2012; Okonya et al. 2013; EAC and USAID 2018). Disaster preparedness and management for these climate impacts in Uganda depends on an accurate analysis and mapping of the vulnerability and susceptibility of communities to risks. This involves geo-referencing, mapping, and livelihood zoning. Undertaking vulnerability analysis is part of early warning systems as articulated by the Ugandan disaster preparedness and response policy (Republic of Uganda 2010).

Vulnerability can also be viewed as susceptibility to harm, which can be defined in terms of population or location (WHO 2013). Location or place in this setting also implies the geography of the area that is vulnerable. Curtis and Jones (1998) point out that place in some respects has an older pedigree in geographies, which aim to provide accounts of the unique social and physical attributes of particular regions. It is also important to appreciate *compositional* and *contextual* effects related to geography and place. Compositional effects arise from the varying distribution of types of people whose individual characteristics influence their health, while contextual effects recognize that the health experience of an individual depends partly on the social and physical environment in the area where they live (Curtis and Jones 1998).

Understanding geography, including the arrangement of health services and the location and nature of environmental exposures, is crucial in assessing the interrelations inherent in many health-related risk exposures (Dummer 2008). Within Africa, the type and magnitude of the health impacts of climate change varies significantly among communities and regions. Variations are due to many factors, such as geographic differences in temperature and precipitation, socioeconomic conditions, the quality of existing health infrastructure, communication capacity, and the underlying disease epidemiology (UNECA 2011). It should also be remembered that climate is not the only factor affecting geographical range and incidence of climate-sensitive health outcomes. Non-climate factors can have strong or even dominant effects, either independently or by modifying climate effects. As such, it is imperative to understand the various causal pathways, from climate change through to health outcomes, to identify opportunities to address the environmental determinants of poor health outcomes (WHO 2013).

Geography, Climate Change and Health Adaptation Planning

Geography studies the physical features of the earth and its atmosphere, and human activity as it affects and is affected by these features, including the distribution of populations and resources and political and economic activities. Given the multitude of issues surrounding climate change science—from root causes to resultant impacts—geography is clearly an elemental factor in the equation. Every aspect of climate change affects or is affected by geography, be it global, regional, or local level (Dangermond and Artz 2010). Therefore, it is important that geography should be considered in health planning, especially in the context of climate change. We now examine how climate change is associated with geography, and how, in turn, geography is being used to inform climate change health planning by referring to specific climate-health events and disasters (see Fig. 13.4).

Health planning is the process of defining health problems, identifying unmet needs and surveying the resources to meet them, establishing priority goals that are realistic and feasible, and projecting administrative action. Health planning is concerned not only with the adequacy, efficacy, and efficiency of health services, but also with those factors of ecology and of social and individual behaviour that affect the health of the individual and community.[1] Planning can also be viewed as the process of setting goals, developing strategies, outlining the implementation arrangements, and allocating resources to achieve those goals. There are many types of planning, including spatial planning and economic planning, among others. Planning for climate change sometimes involves the integration of *adaptation* and *mitigation* in these existing processes, or it can refer to planning processes that are

[1] World Health Organisation Health Strengthening Glossary available online on: www.who.int/healthsystems/hss_glossary/en/index5.html.

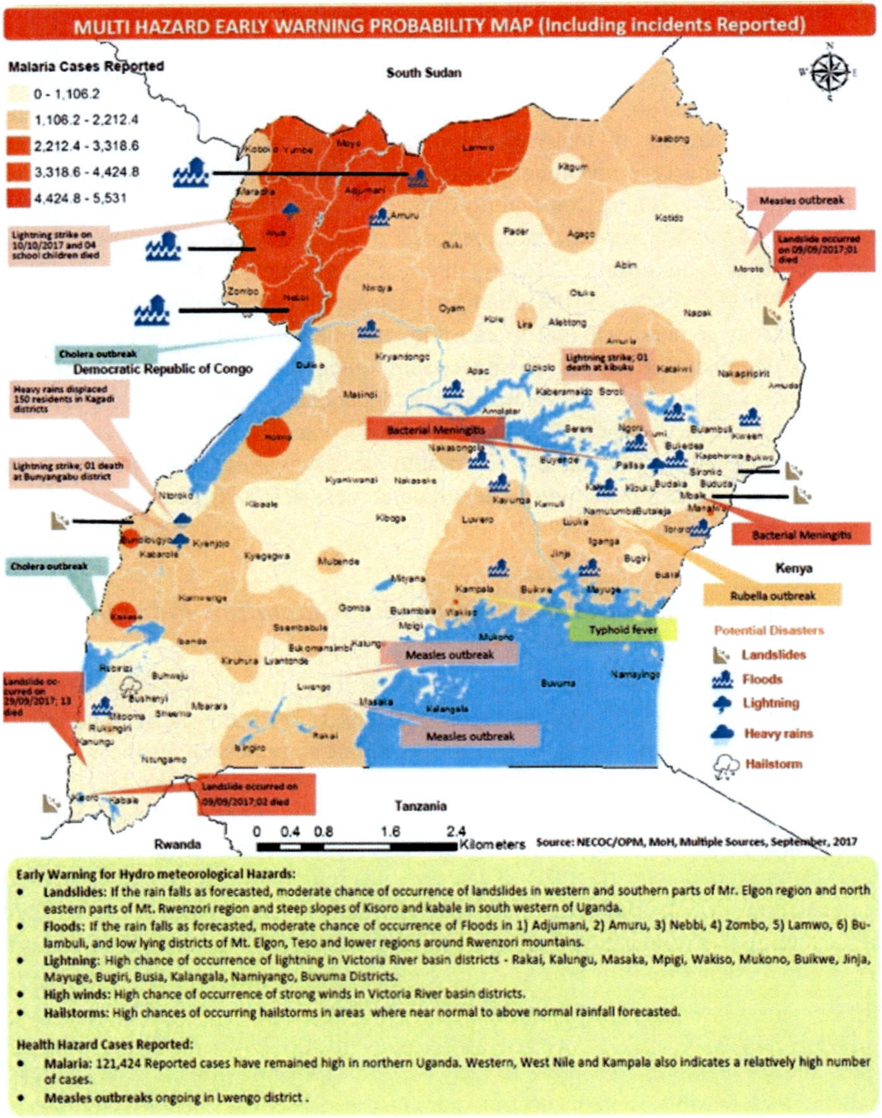

Fig. 13.4 Multi-hazard early warning probability map. (Adopted from Office of the Prime Minister, *Uganda National Integrated Early Warning System Monthly Bulletin, National Emergency Coordination and Operations Centre (NECOC)*, Kampala, Uganda, p. 5)

dedicated specifically to *climate change*, such as the development of a national climate change policy (UNCC: Learn 2014). In terms of mitigation and adaptation, over the long term, mitigation is necessary to avoid continued degradation of the environment determinants of health in the short-to-medium term (i.e. up to the next 20–30 years). Well-planned adaptation measures can avoid much of the projected health impacts (WHO 2015).

Flash and slow-onset floods are perhaps the most common climate-health disasters that are experienced in Uganda. Floods have become more frequent; for example, in 2007 the Teso region of eastern Uganda received the heaviest rainfall experienced in 35 years, affecting 50,000 households and leading to food insecurity (due to loss of harvests). Again in 2010, parts of eastern Uganda experienced unusually heavy and prolonged rains that resulted in flooding (MWE 2014). Areas prone to flooding are predominantly low-lying and located in northern and eastern Uganda and urban centres like Kampala. Figure 13.4 clearly marks out a major geographic flood-prone region associated with the Lake Kyoga and Victoria basins extending along River Nile to the Southern Sudan border. From a similar geographic viewpoint, floods could be properly managed through flood plain mapping and surveys by air and land (Government of Uganda 2010). This type of geographic zoning of flood risk and actual flood occurrence should be integrated into health planning processes at all levels.

Landslides have also been a common disaster in specific geographic regions of Uganda (see Fig. 13.4). For example, tragic landslides occurred in Bududa district in 2010, burying 3 villages and causing death and loss of property, and in 2012 killing 18 people and injuring more than 450 people. In 2011, heavy landslides occurred in neighbouring Bulambuli district, burying several villages (MWE 2014). Areas most affected by landslides are in the Mt. Elgon, Ruwenzori, and Kigezi regions. Although landslides and mudslides are very difficult to predict, their frequency and extent can be estimated by use of information on the area's geology, geomorphology, hydrology, climate, vegetation cover, and traditional knowledge (Government of Uganda 2010). Like floods, the hotspots of landslides in Uganda are strongly associated with the mountainous and highland regions of Rwezori, Elgon and Kigezi. Developing region-specific preparedness plans would reduce health-related and other impacts in these areas.

Drought is another climate-health disaster that requires incorporation into health planning. Many parts of Uganda receive less rainfall than in the past due to climate change and deteriorating regional weather conditions (Government of Uganda 2010). Drought is a normal, recurring feature of climate in most parts of the world. There are three types of drought, namely: meteorological, agricultural and hydrological. Drought can result in malnutrition, dehydration, poor sanitation, and related diseases (UNMA 2017).

The most drought-prone areas in Uganda are the districts in what is called the 'cattle corridor', stretching from western and central to mid-northern and eastern Uganda. In the districts of Karamoja, which are most impacted by drought, starvation, malnutrition, and death of people and livestock has been experienced (Government of Uganda 2010, see Fig. 13.4). The health sector should work closely with the Uganda National Meteorological Authority (UNMA) to forecast drought occurrence in order to inform the districts likely to be affected, so that they can have contingency plan.

As discussed earlier, much of Uganda's disease burden is climate-sensitive and geography related. We tackled malaria and cholera in detail, showing that certain malarial hotspots and cholera belts have emerged. Zinser (2014) indicated that several diseases currently endemic in Uganda will likely increase in prevalence and distribution due to climate change, including malaria and cholera. Interventions need to be targeted to the areas currently affected, as well as those projected to be affected by climate change impacts; for example, heightened surveillance of malaria and cholera in high risk regions, or implementation of indoor residual spraying to prevent malaria epidemics.

The application of geographic information systems (GIS) in health research, planning, monitoring, and evaluation is well established. For example, Shaw et al. (2002) argued that maps are no longer just a way of presenting and summarizing information, but that the advent of geographic information systems (GIS) now permits the mapping process to be an integral part of the research and planning process. As a result, health maps have become widespread due to increased availability of local health data, the development of software solutions, progress in computer capabilities, and a growing interest in geographic inequality in health (Tim et al. 2010). Using GIS allows communities (or sectors like health) to make plans based on the best available scientific information and integrate the needs of both natural and human systems (Dangermond and Artz 2010).

Policy Frameworks and Climate Change Health Responses in Uganda

The health sector-planning regime in Uganda recognizes the impact of climate change on health outcomes. This is clearly illustrated by key policy and strategic documents (MOH 2010a, p. 1). The Health Sector Strategic and Investment Plan (HSSIP) MOH (2010b) under Section 2.3.6 titled; Changing lifestyles and climate change; it is clearly stated that:

> *... Uganda has also experienced the negative consequences of climate change on the health of Uganda for example floods in Eastern Uganda in 2007 resulted in a humanitarian crisis. Higher temperatures and rainfall associated with El Nino may increase the transmission of malaria leading to epidemics in highland areas of Uganda... Better systems for weather forecasting, disease surveillance and public health planning offer some protection for the affected populations. Given the current situation, there is need to emphasize mitigation of adverse effects of climate change...* (p. 21)

In line with international and regional obligations, Uganda developed its 10-year Uganda National Climate Change Policy UNCCP 2015 MWE (2015). The policy's overarching aim is to guide all climate change activities and interventions in the country. In this context, one of the objectives is to support the integration of climate change issues into planning, decision-making, and investments in all sectors and trans-sectoral themes through appropriate institutional arrangements and legal frameworks.

It is important to note that the UNCCP identifies the health sector among the priority sectors that need to plan for climate change. The policy articulates the health policy response and strategic actions to climate change in the health sector:

"to strengthen adaptive mechanisms and enhance early warning systems and adequate preparedness for climate change related diseases.

The prioritization of policies and actions underwent extensive consultations with stakeholders nationally and sub-nationally with both government and non-governmental representatives. Implementation of the specific policies and measures is institutionalized through the UNCCP and led by authorities of the relevant sectors like health (MWE 2015). And this is where the main climate change planning and implementation challenge lies in the health and possibly other sectors. Climate change response is largely viewed as the mandate of the environment sector and it is often not on the priority list of the health sector plans and budgets, both nationally and at the district level. This is in spite of existing climate change mainstreaming guidelines in the sectors. The likely explanation for this climate inaction in the health sector is that this requires additional funding, yet the UNCCP and strategy do not provide it. Consequently, the implementation of key climate change activities, like conducting a vulnerability and adaptation assessment and development of a national health adaptation plan (H-NAP) have not taken off.

Uganda's Climate Change Planning Experiences

Mainstreaming Climate Change in District Health Maps and Plans

Climate change and human health, like other crosscutting concerns such as disability, human rights and gender, are clearly articulated in the Ministry of Health Guidelines to the Local Government Planning Process 2016. The guidelines encourage districts to mainstream climate change activities in all health planning, budgeting, implementation, monitoring and evaluation with the main goal of reducing morbidity and mortality due to climate-related diseases and events.

More specifically, the guide aims to: (1) Ensure health plans are climate-proof (i.e. effectively prepared to respond to climate change impacts); (2) Conduct climate change vulnerability and impact assessment to inform decision-making; (3) Identify opportunities and entry points for integration of climate change mitigation and adaptation (CCMA) measures; and (4) Identify, analyse and integrate options of CCMA into health service delivery.

The climate change health focus and mainstreaming in district plans has been emphasized every year in regional planning meetings in all 14-health regions of Uganda. The health regions are grouped taking into account geographic factors so that unique regional climate concerns like poor rains and famine in Karamoja region or heavy rains and landslides in Mt. Elgon and Mt. Rwenzori regions are handled. While local governments appreciate the importance of planning for climate change, experiences from the planning meetings confirm health system challenges of inadequate financial resources, poor infrastructure and competing priorities of the health sector as limiting factors for mainstreaming climate change activities. Instead, both the national ministry and districts commonly respond when a climate disaster has already occurred resulting in increased fatalities and destruction.

Geography and Climate Change Readiness

Climate change readiness is central to response and planning in the health sector. This includes innovative and practical ways through which a system is prepared to respond to the climate change challenge. Readiness inevitably requires timely weather forecasts and health advisories (e.g. Fig. 13.5 and Box 13.1) from the meteorological services. In this context, the Uganda National Meteorological Authority (UNMA) shares quarterly weather forecasts, which contain health advisories. It is essential to note that forecasts are geographically stratified, grouping districts into agro-ecological zones. These are disseminated to local governments through the District Health Officers to help them plan in advance for heavy rains, drought, and other extreme weather events. Climate services should always ensure that the best available climate information is effectively communicated to people working in sectors such as agriculture, water and health, so that they can develop and evaluate mitigation and adaptation strategies (Jones et al. 2015).

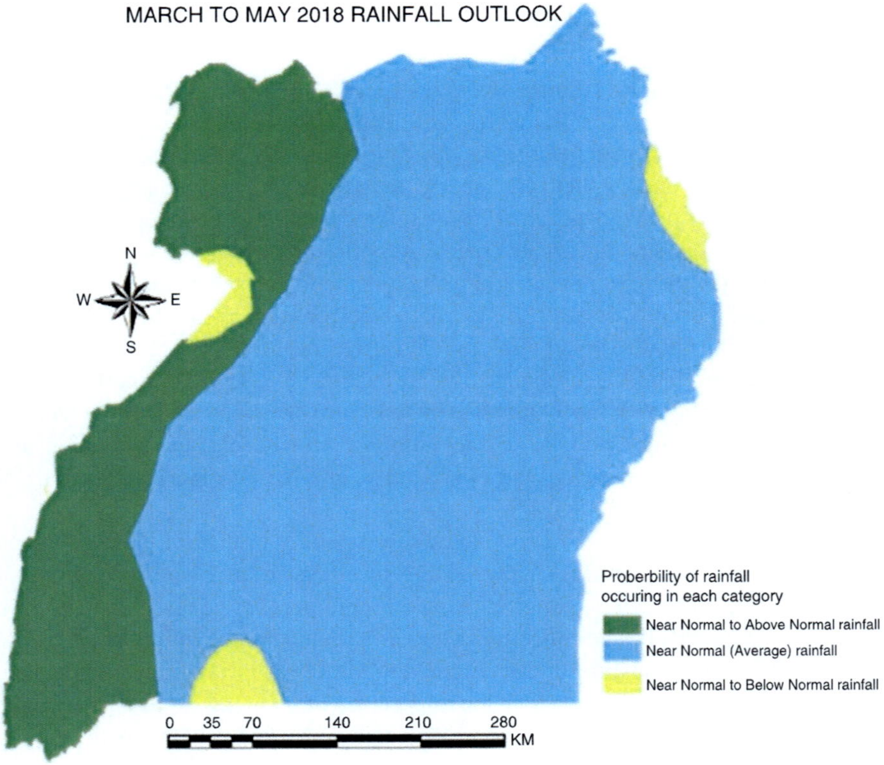

Fig. 13.5 Map for March to May 2018 rainfall outlook over Uganda

Box 13.1 UNMA's Health Advisory Contained in the Seasonal Climate Outlook for March to May 2018 over Uganda
Health advisory

There is need to increase disease surveillance due to expected upsurges of epidemics of diseases that are prevalent during rainy seasons such as malaria, cholera, bilharzia and typhoid. Increases in lower respiratory diseases such as asthma due to humid conditions and allergies from some flowering plants are also expected. Health authorities are therefore advised to be alert and vigilant to equip health units with the necessary drugs to deal with such emergency situations. Therefore, the following measures should be undertaken:

- Intensify health education and awareness campaigns—emphasizing the use of mosquito nets, slashing bushes, disposing open containers, filling up open pits, and draining stagnant water around homesteads in order to reduce breeding places for mosquitoes;
- Frequent health inspection in all communities;
- Increased prevention for air borne diseases and non-communicable illnesses should be improved;
- Improve domestic hygiene and Sanitation around homes and schools to reduce contamination of water, e.g. the use of latrines.

Focused District Visits in Climate Change Disaster Prone Districts

Another geographically oriented strategy aimed at tackling climate change has been focused targeted visits to climate change disaster-prone districts. For example, teams from the Ministry of Health comprising health planners, medical officers, geographers, and environmental health officers have visited the several vulnerable districts to raise climate change awareness and provide guidance in vital climate change planning areas. These districts are hotspots for climate-related disasters like floods and landslides because of their susceptible steep landscapes and floodplains. Meetings are held with members of the District Health Teams (DHTs) to discuss climate change preparedness and response concerns in each district. Emphasis in these meetings is put on climate-informed preparedness plans, emergency systems and community-based disaster and emergency management.

Climate Change and Health Research

Building climate resilience calls for both basic and applied research so as to reduce uncertainty about how local conditions may be affected, gain insight into local solutions and capacities, and build evidence to strengthen decision-making (WHO 2015). In this regard, various research initiatives on climate change have been undertaken in Uganda and East Africa. For example, these include HEALTHY FUTURES (www.healthyfutures.eu), the Indigenous Health Adaptation to Climate

Change (www.ihacc.ca) project, and the East Africa Vulnerability, Impacts, and Adaptation assessment (VIA) undertaken by the East African Community (EAC), with support from the USAID/Kenya and East Africa Planning for Resilience in East Africa through Policy, Adaptation, Research, and Economic Development (PREPARED) Project.[2] These and many others have contributed to the understanding of the complex climate change and health nexus. However, a fundamental gap related to research has been translating the research findings into actual policy and practice. Experience has shown that even when key research findings on climate change and health are shared with decision and policy makers', effective uptake of research recommendations has remained inadequate. There is need for the Ugandan government to invest in climate change research and take charge of this research agenda, including translation of the new knowledge into actual practices. In this regard, there is space for geography and GIS to inform research processes and facilitate programs and policies that target interventions.

Conclusion

This chapter has examined geographic thought in the context of climate change and health adaptation planning in Uganda. The country's diverse geography has strong consequences on its health system particularly the disease burden, climate variability and change and related vulnerabilities and impacts. Major diseases like malaria and cholera are climate-sensitive and geography-related. We have illustrated that Uganda's health system is highly sensitive to climate change impacts, but also that its adaptive capacity is inadequate. The existing national policy framework is sound enough to tackle the climate change challenge. However, the fundamental barrier has been lack of comprehensive health vulnerability and adaptation assessment to support evidence-based health adaptation planning and strategy development. Furthermore, there is a question of where additional financing for climate change activities will be accessed. The climate change policy structure should urgently devise financing mechanisms at national and sub national levels if the policy is to have the desired impact. The modest experiences in climate change health adaptation planning shared in this chapter can inform future steps like vulnerability assessment and H-NAP. A multi-sector approach emphasizing linkages and collaboration in planning, implementation, monitoring and evaluation is highly recommended. Given the discussion in this chapter, it is also inevitable that geographical thought should inform health climate change adaptation planning. Moreover, as Dangermond and Artz (2010) pointed out, making decisions based on geography is basic to human thinking.

[2] Vulnerability, Impacts and Adaptation Assessment in the East African Region, Chapter 5: Health, sanitation and human settlements, available on: http://files.constantcontact.com/a39f63e2501/ddca683e-f575-4a46-9004-d4175945b081.pdf, viewed 8/5/2018.

References

Barnett, J., & Adger, W. N. (2007). Climate change, human security, and violent conflict. *Political Geography, 26*, 639–655. https://doi.org/10.1016/j.polgeo.2007.03.003.

Connor, S. J., et al. (2006). Using climate information for improved health in Africa: Relevance, constraints and opportunities. In *United Nations Economic Commission for Africa 2011, climate change and health across Africa, Working paper* (p. 20). Addis Ababa: African Climate Policy Centre.

Curtis, S., & Jones (1998). Is there a place for geography in the analysis of health inequality? *Sociology of Health & Illness, 20*(5), 645–672. ISSN 0141-9889.

Dangermond, J., & Artz, M. (2010). *Climate change is a geographic problem*. Redlands, CA: ESRI.

Dummer, T. J. B. (2008). Health geography: Supporting public health policy and planning. *CMAJ, 178*(9), 1177–1180.

EAC and USAID (East African Community and United States Agency for International Development). (2018). *Climate change vulnerability and adaptation in East Africa—Health, sanitation and human settlements, Fact sheet*, Arusha.

Government of Uganda. (2010). *The national policy for disaster preparedness and management*. Kampala: Directorate of Relief, Disaster Preparedness & Refugees, Office of Prime Minister.

Hartter, J., Stampone, M. D., Ryan, S. J., Kirner, K., Chapman, C. A., & Goldman, A. (2012). Patterns and perceptions of climate change in a biodiversity conservation hotspot. *PLoS One, 7*. https://doi.org/10.1371/journal.pone.0032408.

Hepworth, N. D. (2010). *Climate change vulnerability and adaptation preparedness in Uganda*. Nairobi: Heinrich Böll Foundation.

Hussein, M. A. (2011). Climate change impacts on East Africa. In W. L. Filho (Ed.), *The economic, social and political elements of climate change* (pp. 589–601). Berlin: Springer-Verlag.

IOM—Institute of Medicine. (2001). *Health and behaviour*. Washington, DC: National Academies Press.

IPCC. (2007). Summary for policy makers. In S. Solomon et al. (Eds.), *Climate change 2007: The physical science basis. Contribution of working group I to the fourth assessment report of the Intergovernmental Panel on Climate Change*. Cambridge: Cambridge University Press.

Jones, L., Carabine, E., Roux, J. P., & Turner, T. (2015). *Promoting the use of climate information to achieve long-term development objectives in sub-Saharan Africa: Results from the future climate for Africa scoping phase*. Climate, Development and Knowledge Network (CDNK).

Kwesiga, B., et al. (2018). A prolonged, community-wide cholera outbreak associated with drinking water contaminated with sewage in Kasese district, western Uganda. *BMC Public Health, 18*. https://doi.org/10.1186/s12889-017-4589-9.

Magrath, J. (2008). Turning up the heat: Climate change and poverty in Uganda. *Oxfam Policy and Practice: Agriculture, Food and Land, 8*, 96–125.

MFPED. (2015). *Second National Development Plan (NDPII) 2015/2016–2019/20*. Kampala: Republic of Uganda.

Ministry of Health. (2010a). *The second National Health Policy*. Kampala: Government of Uganda.

Ministry of Health. (2010b). *Health sector strategic and investment plan*. Kampala: Government of Uganda.

Ministry of Health. (2014). *Uganda malaria reduction strategic plan*. Kampala: Government of Uganda.

Ministry of Health. (2016a). *Guidelines to the local government planning process*. Kampala: Government of Uganda.

Ministry of Health. (2016b). *Annual health sector performance report*. Kampala: Government of Uganda.

Ministry of Health. (2017). *Prevention and control of cholera: Operational guidelines for the national and district health workers and planners*. Kampala: Government of Uganda.

Ministry of Water and Environment (MWE). (2014). *Guidelines for the integration of climate change in sector plans and budgets*. Republic of Uganda, Kampala.

Ministry of Water and Environment. (2015). *Uganda's intended nationally determined contributions (INDC)*. Kampala: Ministry of Water and Environment.

Namanya, D. B. (2009). *An assessment of the impact of climate change on the health sector in Uganda: A case of malaria and cholera epidemics and how to improve planning for effective preparedness and response* (pp. 1–32). Nakasero: Ministry of Health.

Okonya, J. S., Syndikus, K., & Kroschel, J. (2013). Farmers' perception of and coping strategies to climate change: Evidence from six agro-ecological zones of Uganda. *The Journal of Agricultural Science, 5*, p252.

Republic of Uganda. (2010). *The national policy for disaster preparedness and management, directorate of relief, disaster preparedness and refugees*. Kampala: Office of the Prime Minister.

Shaw, M., Dorling, D., & Mitchell, R. (2002). *Health, place and society*. Essex: Pearson Educational Ltd.

Tim, B., McLafferty, S., & Moon, G. (Eds.). (2010). *A companion to health and medical geography*. Sussex: Blackwell Publishing Ltd.

Tunstall, H. V. Z., Shaw, M., & Dorling, D. (2004). Places and health. *Journal of Epidemiology and Community Health, 58*, 6–10.

UNCC: Learn. (2014). *Introduction to planning for climate change*. The One United Nations Climate Change Learning Partnership.

United Nations Economic Commission for Africa. (2011). *Climate change and health across Africa, working paper 20*. Addis Ababa: African Climate Policy Centre.

UNMA (Uganda National Meteorological Authority). (2017). *Frequently asked question on weather and climate*. Kampala: UNMA.

Wang, H., & Horton, R. (2015). Tackling climate change: The greatest opportunity for global health. *Lancet, 386*, 7–13.

WHO. (2013). *Protecting health from climate change: Vulnerability and adaptation assessment*. Geneva: WHO.

WHO. (2015). *Operational framework for building climate resilient health systems*. Geneva: WHO.

Zinser, K. (2014). *An overview of climate change and health in Uganda*. Washington, DC: USAID, Tetra Tech ARD.

Chapter 14
Mobile Health Geographies: A Case from Zimbabwe

Liberty Makacha and Prestige Tatenda Makanga

Abstract Mobile health (mHealth) commonly intersects mobile technology and healthcare practices. Geographically enabled mobile health is an evolving subset of mHealth practices that accounts for how geography mediates health outcomes. The domain of both mHealth and geo-enabled mHealth is varied in both scope and interpretation. This chapter explores some of the varied interpretations of mHealth and geographically enabled mobile health. Emerging trends and trajectories in geographically enabled mHealth as well as barriers to implementation of geo-enabled mobile health practices in Africa are also discussed. These are explored through two cases: Mapping Outcomes for Mothers (MOM) and RoadMApp, both aimed at improving maternal health outcomes in low resourced settings. At the time of writing this chapter, both mHealth applications were at the prototype phase with plans for testing them in Zimbabwe, Kenya, the Gambia, and Mozambique. The chapter concludes by discussing how these and similar emerging technologies possess incredible potential for targeting health interventions. However, the acceptability, clinical utility and access to geo-enabled mHealth tools are yet to have demonstrable impact on health policy, especially in Africa.

Keywords mHealth · Geographically enabled mobile health · eHealth · Public health · Health geography

L. Makacha (✉) · P. T. Makanga
Place Alert Labs (PALs), Department of Surveying and Geomatics, Midlands State University, Gweru, Zimbabwe
e-mail: makachal@staff.msu.ac.zw

P. T. Makanga (ed.), *Practicing Health Geography*, Global Perspectives on Health Geography, https://doi.org/10.1007/978-3-030-63471-1_14

Mobile Health Geographies

Mobile Health (mHealth)

mHealth is a global healthcare technology concept believed to have been first coined by Robert S.H. Istepanian in 2003 (Woodward 2016). The concept of mHealth underpins the use of mobile devices in healthcare provision. The term has also been used to define other varied forms of wireless technologies aiding health practice (Jill et al. 2016). Lately, mHealth technologies have found space in the literature with a good presence especially in technology reports. The presence of these technologies in the literature from the lesser developed regions of the world is on an increase. This has been attributed to the sparse and over-stretched nature of healthcare provisions especially in the last decade (Siddique et al. 2017). Current trends leading to the increased usage of these technologies are mainly a function of the ubiquitous uptake of mobile phone usage especially in the first quarter of the twenty-first century (Vital-Wave-Consulting 2009).

While a standardized definition for mHealth is still absent in the literature, indicating the multiplicity of perspectives around the function of mobile technology in improving health states, several definitions have been coined for the same. Table 14.1 summarizes some of the varying interpretations of mHealth. It is important to note that although the definitions and/or interpretations of mHealth are somewhat varied, mobile technologies form an important central component for mHealth.

This chapter describes some of the key characteristics of mHealth. It also acknowledges the role of mHealth tools and applications in self-care and/or self-managed health care practices.

In as much as mHealth applications have gained recognition in the past decade, a variant of these applications, geo-mHealth applications, are increasingly gaining traction, especially in public health practices. There is generally much to be optimistic about geo-enabled m-health as access to location enabled mobile technologies continue to improve.

Geographically Enabled mHealth

Geographically enabled mHealth (Geo-mHealth) is a specialized set of mHealth tools that use geographically referenced data and maps as the basis for bringing context specific intelligence into health decisions and interventions. They are a derived subset of mHealth, bearing a lot of resemblance to mHealth applications in form and function, with an added capability of geographical precision and awareness. Geo-mHealth is mobile health (mHealth) technologies delivering healthcare services, overcoming spatio-temporal limitations often characterizing m-Health applications. In some literature this is referred to as location-based health information services (Boulos 2003), though in the context of this chapter, the same information services are specifically implemented through the use of mobile technology.

Table 14.1 Definitions and key features of mHealth

Definition	Key features in definition	References
A redefinition of telemedicine	Definition speaks more to some of the key application domains of mHealth	Free et al. (2010)
May be interpreted as medicine practiced at a distance, including new mobility and wireless communication services to reshape the future of global health systems	Definition also emphasizes some of the key technologies embedded in mHealth Definition also speaks to the utilities embedded in mHealth tools	Istepanian et al. (2004)
Medical and public health practices supported by mobile phones, patient monitoring devices and personal digital assistants among other wireless technologies	This definition contextually narrows down to mHealth utilities in clinical and public health practice only This definition principally lacks an explanation of what the technology really is outside the technology's utilities	WHO (2011)
The use of mobile technologies in the practice of medicine and the transmission of health-related information over mobile networks to improve communication between patients and clinicians	There is a huge degree of overlap between this definition and the definition by WHO (2011) The definition emphasizes the communication interface between the technology and the physician rather than the mHealth technologies themselves	Wickramasinghe (2016) as quoted in AHIMA Guide 2013

Geo-mHealth will potentially address some of the emerging problems of health services delivery, including the need to provide direct and timely access to health services, as well as to potentially scale down on health-related costs. Leveraging integrated geospatial information and technologies has the potential to transform the way health care providers use sophisticated mobile technologies to gain insights into clinical and other data repositories allowing passing down well-informed health decisions at the individual level.

While it is important for caregivers to know the clinical history of the patients they treat, information concerning the social and environmental contexts within which those patients live is known to be an important determinant of their health outcomes. Geography is a strong modifier of health outcomes, necessitating factoring in of key socio-geographical variables in epidemiological assessments (see Chap. 4). However, a big challenge remains especially in Africa to find and push location-specific strategies to users of mHealth tools based on their location and contexts (Brennan and Friede 2001). The integration of location intelligence into mHealth systems is still at the norming phase, especially in Africa to the extent that there is very little explicit mention of the practice of Geo-mHealth in African literature.

Evidence for the uptake of Geo-mHealth in sub-Saharan Africa is somewhat porous. The eastern region of sub-Saharan Africa had a relatively higher uptake of mHealth applications (17 countries with 287 mHealth programs), while the western region had 16 countries with 145 mHealth programs (Seohyun et al. 2017). However, it remains unclear how many of these mHealth applications are geographically intelligent. These trends are expected to grow exponentially in the coming decade to 2030.

Lately, efforts to make these systems geographically aware have been a core push in technology reviews. However, current infrastructure largely relies on global open mapping services such as Google Maps. The lack of local spatial data infrastructures in developing countries in Africa has greatly impeded growth of these technologies. However, new policy priorities in global health related to granularizing health records and using the same for informing targeted public health interventions present unique opportunities for Geo-mHealth to develop and thrive. The following section describes two examples that illustrate this unique opportunity.

Mobile Health Geographies: Case Studies

The two cases presented in this chapter are examples of Geo-mHealth tools that are in the prototype phase at the time of writing. These cases illustrate the value that geography brings to mHealth as summarized in Table 14.2. These Geo-mHealth cases have been selected because of their contextual relevance to the African setting.

Mapping Outcomes for Mothers (MOM)

The Mapping Outcomes for Mothers (MOM) Geo-mHealth application incorporated community-specific determinants of health, as well as issues related to geographical access (Firoz et al. 2017). Data on adverse maternal outcomes and their potential determinants were used to generate spatial models that described community-level risk and resilience related to maternal health in regions of Southern Mozambique (see Chap. 4). This evidence was used to inform the location intelligence behind MOM. MOM is intended for use by the community health worker during her routine visits to pregnant women. It is a place aware application that 'knows' the nonclinical personal and community-level determinants that elevate risk for an adverse maternal outcome. Some of these determinants include, women's financial autonomy in pregnancy, access to help from family in the event of a pregnancy related need, distance from health facility, etc.

MOM considers the location of the pregnant woman, her personal socio-demographic characteristics and the character of here immediate environment/community to generate a risk profile for the woman. This Risk score is determined through spatial algorithms that run on the phone (Fig. 14.1). The community health worker then

Table 14.2 Key features of the MOM and RoadMApp mhealth applications

Geo-enabled mHealth application	Key features of the application	Notable/perceived impact on public health practice
Mapping outcomes for mothers (MOM)	MOM utilizes geographic methods to describe the contextual social and environmental determinants that elevate risk of adverse maternal outcomes in Africa	– Community health workers (CHWs) are presented with a list of risk factors and recommendations relevant to the pregnant woman's personal circumstances and those of the community where she lives – Some of the identified risk factors factored into the mobile application include quality of roads, marital status, isolation (as measured by travel distance to the nearest major road), women's financial autonomy in pregnancy as well as family support (both financial and transport)
RoadMApp	RoadMApp is a geographically intelligent mobile application designed to link pregnant women to locally available transport resources as well as providing a micro-savings scheme for financing transport related to their maternal health needs	The pregnant woman, health facility and drivers (pooled locally from the community) utilize a common platform to facilitate the woman's access to maternal care services Three components that constitute the RoadMApp system • Communicating travel risk (travel time based on near real-time travel and weather conditions) • Facilitating saving for transport during the tenure of pregnancy for seeking antenatal and emergency pregnancy care • A ride-sharing model to facilitate for transportation to health facilities

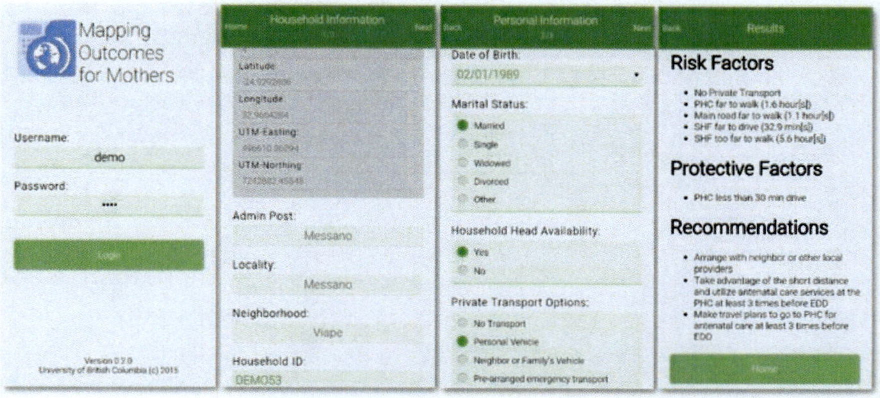

Fig. 14.1 Screenshots from the MOM app

informs the woman of these risk factors. In addition, MOM uses evidence of what is known to work in other communities to promote maternal resilience by generating a list of recommendations that are related to the woman's risk profile. This utilizes the mobile phone as the means for bridging social capital from one community to the next.

While MOM was developed using data collected in a clinical trial, scaling up the innovation and sustaining its use at scale will require support from a broader data and technology infrastructure base (Fig. 14.2). This will have to include National demographic databases, and a rich spatial data infrastructure to form the basis for generating geographic intelligence on the place specific character of illness, health, and all associated determinants. The use of the app will also require to be extended to health care administrators and planners to monitor place specific maternal health trends, utilizing MOM as both an infield health decision aid, as well as a data capture/surveillance toolkit.

RoadMApp

Poor access to maternal health services has been known to impact negatively on the health system delivery especially in low resourced settings. The RoadMApp intervention is one geographically enabled mobile health application which improves maternal health outcomes by facilitating access to pregnancy related care using a mobile platform. RoadMApp is a near real-time location sensitive mobile technology

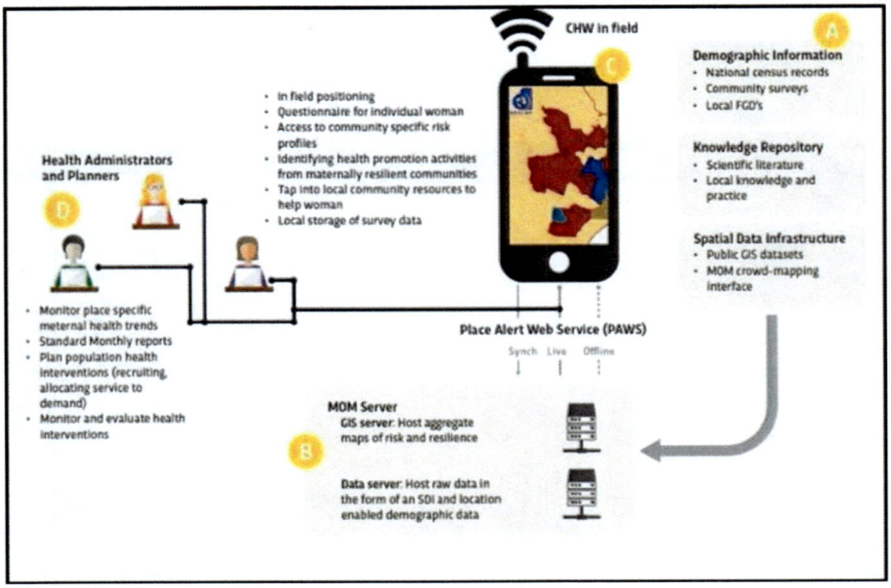

Fig. 14.2 Entity interactions—the Mapping Outcomes for Mothers. (Figure taken from PRE-EMPT 2018)

solution built upon a novel spatio-temporal model of access to maternal health care that will inform pregnant women about travel times to, and potential isolation from health facilities. This mHealth tool accounts for roads and weather conditions in near real time, linking women to locally available transport resources to facilitate their transit to care facilities (Fig. 14.3). The innovation also translates culturally established micro-savings traditions to finance access to care especially in emergencies or where transport financing cannot be met through subsidized government or other pooled resources.

This is the first time that geographical access to maternal health services, and the transport and financial resources to support it, have been addressed through a single location-based platform. The result is that women will not only be informed of travel risk but pointed to local resources to mediate it. By masking much of the complexity behind spatial models of access to care, RoadMApp will facilitate context specific pregnancy decision making at the individual rather than population level resulting in more precise and efficient responses to pregnancy risk. RoadMApp will have the greatest impact on improving access for marginalized populations (both geographically and socially) by linking them to local transport and financial resources.

RoadMApp will provide solutions to the first and second delay for accessing maternal care by addressing the barriers induced by long travel times, weather, scarcity, and cost of transport to both health facilities and maternity waiting homes. The RoadMApp intervention utilizes a ride-sharing model that links women to locally

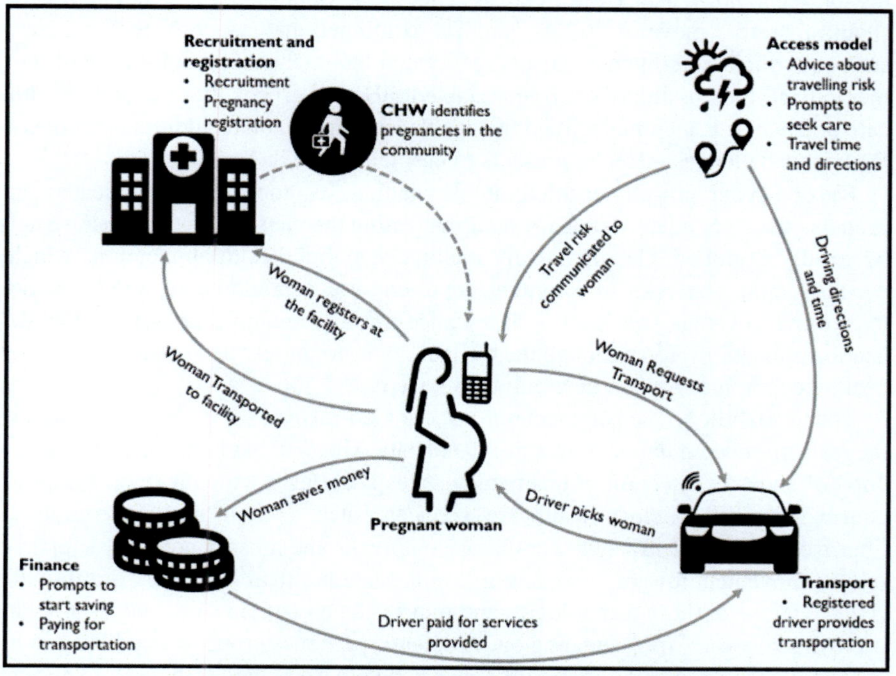

Fig. 14.3 RoadMApp intervention diagram. (Adapted from the RoadMApp protocol)

available transport crowd sourced through the mobile platform. Transportation costs are financed through a community led "mobile money" savings platform that the community and pregnant women contribute to. RoadMApp, however, relies on the availability of externally operated mobile services which limits its utility in poor network environments.

The Future of Geo-mHealth

There is generally a serious need to revisit current understandings and interpretations of geographically aware mHealth practices, envision how these may be superseded with a new paradigm shift in the status quo of technology and health systems, leading to more successful future m-Health systems. The world over, we are currently witnessing an exponential growth in remote monitoring and mobile applications for healthcare. These solutions are all designed to ultimately enable the health service consumer to enjoy better healthcare delivery and/or wellness (Ruwini et al. 2017).

The cloud provides a means for reducing the information technology footprint within a given health system, both public health and clinical practice, while at the same time opening the door to delivering data and applications to end users more efficiently and effectively. However, in Africa there is a noticeable gap in mobile health geographies mainly because Africa has largely been left behind in the information technology gap. Health care practices that want to put patient/client care should therefore move to fill this gap. The continued maturation of technologies, such as artificial intelligence, virtual reality and block-chain technologies will further expand the possibilities for geo-enabled mHealth especially in clinical health care practice. It is incumbent upon the health care profession to address proactively the many challenges mHealth presents to best maximize its benefits.

Beyond geographically enabled mobile health as just tools to communicate map products there are other avenues of communicating the map product which have not been fully exploited. Geographically intelligent mobile health applications which use messaging platforms to communicate to end users should be explored in a bid to improve coverage and level of acceptance of the communications from mobile health applications. Even though the technology is not purely new, its use in mobile health geographies has not been fully embraced.

The availability of sensing technologies in Geo-mHealth is envisioned to allow the real-time accumulation of health-related data. This will likely increase the likelihood of an early detection of emergency states enabling a wide range of versatile, convenient, and efficient healthcare services, such as real-time interventions, improved abatement strategies for disease outbreaks and most importantly supporting the movement toward "maintaining wellness rather than sick care" (Latif et al. 2017). Uptake of these technologies and improved perceptions of the odds for their use will likely see Africa standing out to defend epidemics proneness and fair better against other regions, considering that her people have tended to be adaptive even without such technologies.

Conclusion

The chapter discussed context specific mobile health geographies as well as the future of geographically enabled m-Health. We presented future directions to potentially shape and transform healthcare services in Africa using geographically enabled mHealth services. However, in as much as Geo-mHealth is a promising technology for the betterment of public health practice in Africa, it is surrounded by enormous challenges. The technology will likely benefit from current advances in ICT, especially cloud computing facilities. A robust and secure Geo-mHealth could be an important and inevitable future development for Africa's people in delivering efficient and reliable health care. Geo-mHealth is a potential enabler in making some elements of health care faster, better, more accessible, and cheaper. The gap in uptake and acceptability between Africa and the first world remains an issue of concern both for policy implementers and the health system in general.

References

Boulos, M. (2003). Location-based health information services: A new paradigm in personalised information delivery. *International Journal of Health Geographics, 10*(2), 2.

Brennan, P., & Friede, A. (2001). Public health and consumer uses of health information. In E. Shortliffe & L. Perreault (Eds.), *Medical informatics—Computer applications in health care and biomedicine* (pp. 397–420). New York: Springer-Verlag.

Firoz, T., Makanga, T., Nathan, H. L., Payne, B., & Magee, L. A. (2017). Reverse innovation in maternal health. *Obstetric Medicine, 10*, 113–119.

Free, C., Phillips, G., Felix, L., Galli, L., Patel, V., & Edwards, P. (2010). The effectiveness of M-health technologies for improving health and health services: A systematic review protocol. *BMC Research Notes, 3*, 250.

Istepanian, R., Jovanov, E., & Zhang, Y. (2004). Introduction to the special section on m-Health: Beyond seamless mobility and global wireless health-care connectivity. *IEEE Transactions on Information Technology in Biomedicine, 8*(4), 405–414.

Jill, F., Faatiema, S., Leila, H. A., & Tania, S. D. (2016). The impact of mHealth interventions on health systems: A systematic review protocol. *Systematic Reviews, 5*, 200.

Latif, S., Rana, R., Qadir, J., Ali, A., Imran, M. A., & Younis, M. S. (2017). Mobile health in the developing world: Review of literature and lessons from a case study. *IEEE Access, 5*, 2169–3536. https://doi.org/10.1109/ACCESS.2017.2710800.

PRE-EMPT, U. C. (2018, February 27). *Pre-eclampsia and eclampsia monitoring, prevention and treatment*. Retrieved May 1, 2018 from https://pre-empt.bcchr.ca/monitoring/mapping-outcomes-mothers-mom.

Ruwini, E., Stranieri, A., & Wickramasinghe, N. (2017). A taxonomy for mHealth. In *Handbook of research on healthcare administration and management*. Hershey, PA: IGI Global.

Seohyun, L., Yoon-min, C., & Sun-Young, K. (2017). Mapping mHealth (mobile health) and mobile penetrations in sub-Saharan Africa for strategic regional collaboration in mHealth scale-up: An application of exploratory spatial data analysis. *Globalization and Health, 13*, 63.

Siddique, L., Rajib, R., Junaid, Q., Anwaar, A., Muhammad, A. I., & Shahzad, Y. (2017). Mobile health in the developing world: Review of literature and lessons from a case study. *IEEE Access, 5*, 11540–11556.

Vital-Wave-Consulting. (2009). *mHealth for development: The opportunity of mobile technology for healthcare in the developing world.* Washington, DC and Berkshire: UN Foundation-Vodafone Foundation Partnership.

WHO Global Observatory for eHealth. (2011). *mHealth new horizons for health through mobile technologies.* Geneva: Global Observatory for e-Health of the World Health Organization. Retrieved April 25, 2018 from http://www.who.int/goe/publications/goe_mhealth_web.pdf.

Wickramasinghe, N. (2016). *Handbook of research on healthcare administration and management.* Hershey, PA: IGI Global.

Woodward, B. (2016). *m-Health: Fundamentals and applications.* Piscataway, NJ: The Institute of Electrical and Electronics Engineers, Inc..

Index

P. T. Makanga (ed.), *Practicing Health Geography*, Global Perspectives on Health Geography, https://doi.org/10.1007/978-3-030-63471-1